BIOLOGY

W R Pickering

Oxford University Press 1996

Oxford University Press,
Great Clarendon Street, Oxford OX2 6DP

Oxford New York
Athens Auckland Bangkok Bogota Bombay
Buenos Aires Calcutta Cape Town Dar es Salaam Delhi
Florence Hong Kong Istanbul Karachi
Kuala Lumpur Madras Madrid Melbourne
Mexico City Nairobi Paris Singapore
Taipei Tokyo Toronto
and associated companies in
Berlin Ibadan

Oxford is a trade mark of Oxford University Press

© **W. R. Pickering**

All rights reserved. This publication may be reproduced, stored or transmitted, in any forms or by any means, except in accordance with the terms of licences issued by the Copyright Licensing Agency, or except for fair dealing for the purposes of research or private study, or criticism or review, as permitted under the Copyright, Designs and Patents Act, 1988. Enquiries concerning reproduction outside those terms should be addressed to the Permissions Department, Oxford University Press.

First published 1996

ISBN 0 19 914301 3 (School edition)
      0 19 914302 1 (Bookshop edition)

Typesetting, design and illustration by Hardlines, Charlbury, Oxford
Printed in Great Britain

CONTENTS

LIFE PROCESSES
Aerobic respiration	5
Human body systems	6
Angiosperm adaptations	7
Plant and animal cells	8
Diffusion, osmosis and active transport	9
Enzymes control biological processes	10

HUMAN PHYSIOLOGY
An ideal human diet	11
Malnutrition	12
Digestion, absorption and processing of foods	13
Mammalian circulation	14
Blood structure and function	15
Disease may have a number of causes	16
Food preservation	17
Natural defence systems of the body	18
Antibodies are specific protein molecules	19
Abuse of drugs	20
Neurones (nerve cells) carry information through the nervous system	21
The endocrine system	22
The eye is a sense organ	23
The brain is an integrator	24
Blood sugar regulation	25
Excretion	26
Kidney structure and function	27
Kidney failure: dialysis and transplants	28
Temperature control in endothermic organisms	29
Muscle-bone machines	30
Human reproduction	31
The menstrual cycle	32
The placenta	33

PLANT PHYSIOLOGY
Plants make food by photosynthesis	34
Transport systems in plants	35
Water movement through a plant	36
Hormones and minerals affect plant growth and development	37
Flower structure is adapted to pollination	38
Fertilisation, fruits and seed dispersal	39

VARIATION, INHERITANCE AND EVOLUTION
Keys and classification	40
Naming and classifying living organisms	41
Cell division and the human life cycle	42
DNA and chromosomes	43
DNA, genes and proteins	44
Cystic fibrosis and monohybrid inheritance	45
Sex linkage and the inheritance of sex	46
Variation, natural selection and evolution	47

BIOTECHNOLOGY AND GENETIC ENGINEERING
Genetic engineering (recombinant DNA technology)	48
Gene transfer can promote desirable characteristics	49
Cloning	50
Bioreactors/fermenters	51
Penicillin is an antibiotic	52
Bacteria and food production	53
Anaerobic respiration	54

ECOLOGY AND ECOSYSTEMS
Ecology	55
Factors affecting population growth	56
Human population growth	57
Pollution of the atmosphere	58
Pollution of water	59
Saprotrophs cause decay	60
Treatment of sewage	61
Ecological pyramids	62
Biological pest control	63
Cycling of nutrients	64
Managing ecosystems: fish farming	65
Managing ecosystems: horticulture	66
Managing ecosystems: animal husbandry	67
Design an experiment	68
Dealing with data	69
Graphical representation	70

INDEX 71

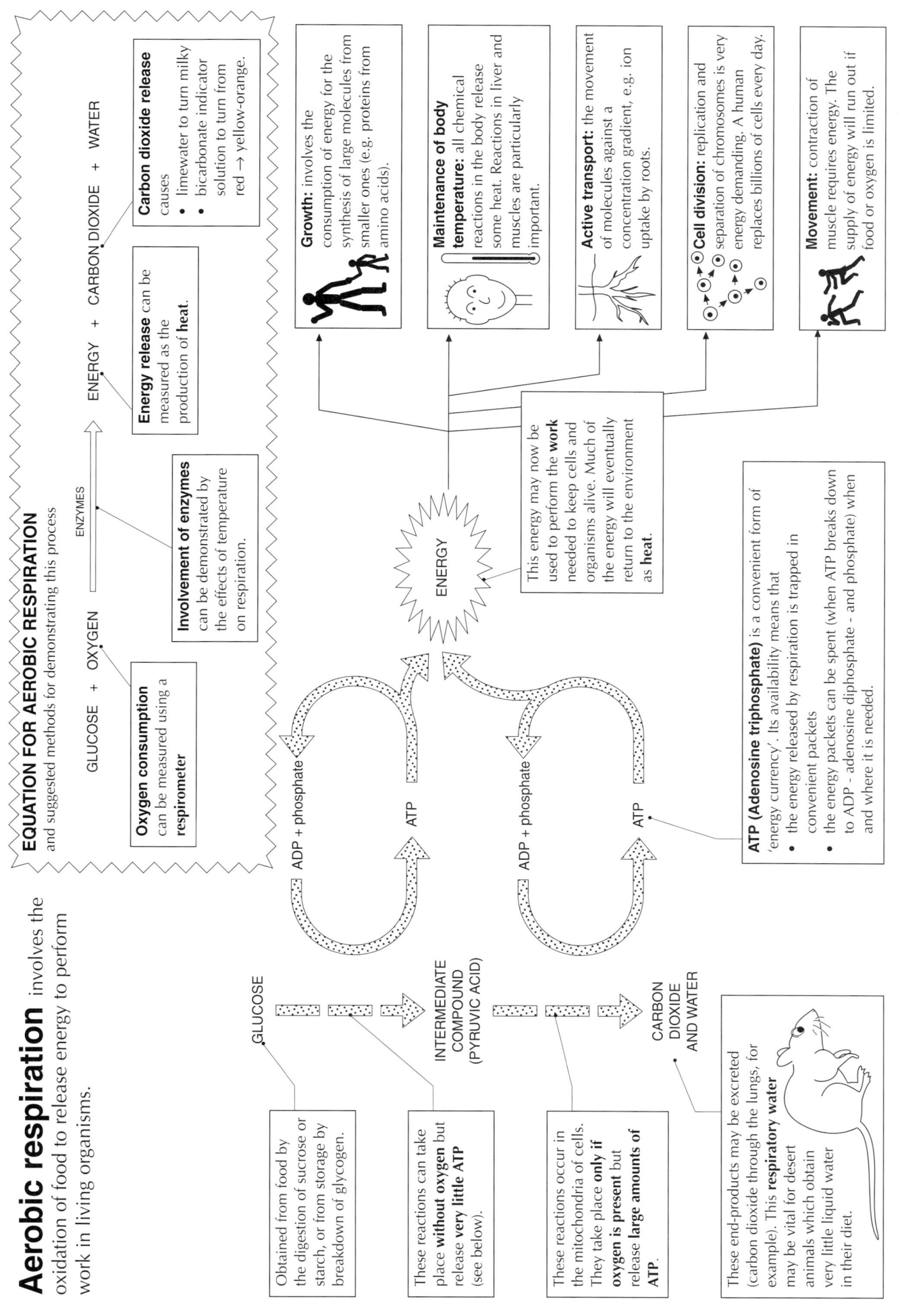

Human body systems

all play a part in homeostasis.

HUMANS ARE MAMMALS.

Humans
- breathe through lungs
- have a diaphragm separating chest from abdomen
- fertilise eggs and develop their young inside the body of the female
- bear live young
- feed their young on milk from specialised mammary glands
- have some body hair, which may once have helped to make humans endothermic

** **Humans are extremely successful because they are homeostatic (can keep a 'steady state') and thus independent of their environment**

The circulatory system is composed of two subsystems. The **cardiovascular system** (heart, blood vessels and blood) is the transportation system responsible for distribution of many solutes, and the **lymphatic system** (lymph vessels and lymph) is responsible for return of tissue fluid to blood and for defence against diseases.

The respiratory system consists of the **lungs** and **air passageways** and maintains optimum concentrations of carbon dioxide and oxygen in the tissues.

The urinary system comprises the **kidneys**, **bladder** and associated **ducts**. These produce urine and remove it from the body. They remove toxins from, and maintain optimum solute concentrations in the blood.

The digestive system comprises **mouth**, **oesophagus**, **stomach**, **intestines** and accessory organs (principally **liver** and **pancreas**). These organs ingest food, break it down mechanically and chemically, and absorb nutrients so maintaining optimum concentrations of fuel molecules and raw materials for syntheses.

The nervous system is made up of **brain**, **spinal cord**, **sense organs** and **peripheral nerves**. Conducts impulses responsible for integration of other systems. The principal regulatory system.

The muscular system comprises **skeletal**, **cardiac** and **smooth muscle**. Responsible for the locomotion, movement of body parts, pumping of blood and internal movement of other material. Maintains blood pressure, responsible for formation of tissue fluid.

The skeletal system consists of **bones** and **cartilage**. Supports the body, protects soft tissues, site of calcium storage and blood cell synthesis.

The endocrine system consists of the **ductless glands**, many of which are under the influence of the pituitary gland. These regulate many body functions, and help to keep a constant composition of the blood.

The reproductive systems comprise **testes** (male), **ovaries** (female) and associated structures. Gamete production and transfer, and maintenance of secondary sexual characteristics.

Skin, **hair**, **nails** and **sweat glands** protect body against infection and dehydration, help control body temperature and receive stimuli such as pressure.

Angiosperm adaptations
- more than 80% of all plant species are angiosperms (i.e. plants with enclosed seeds).

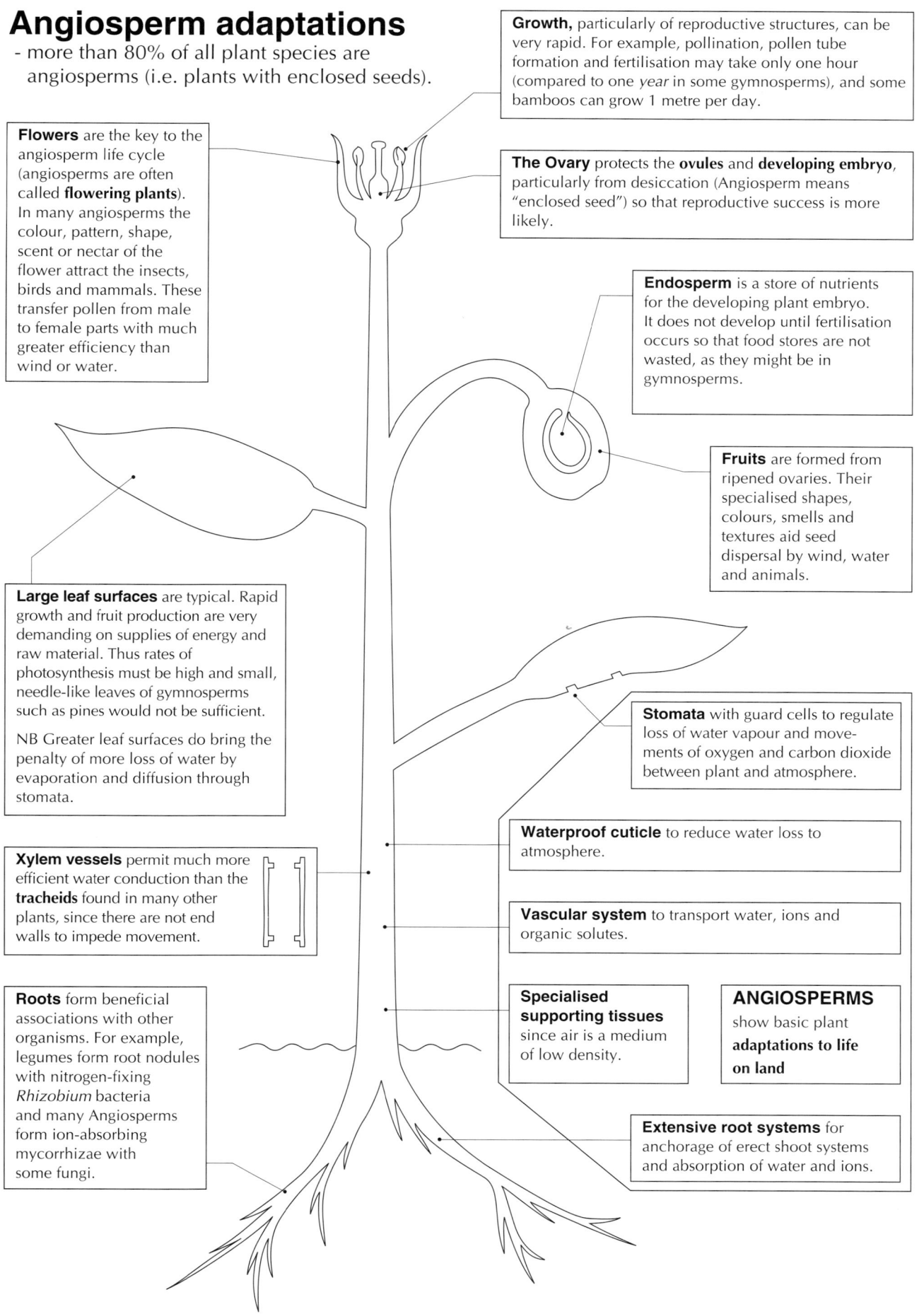

Growth, particularly of reproductive structures, can be very rapid. For example, pollination, pollen tube formation and fertilisation may take only one hour (compared to one *year* in some gymnosperms), and some bamboos can grow 1 metre per day.

Flowers are the key to the angiosperm life cycle (angiosperms are often called **flowering plants**). In many angiosperms the colour, pattern, shape, scent or nectar of the flower attract the insects, birds and mammals. These transfer pollen from male to female parts with much greater efficiency than wind or water.

The Ovary protects the **ovules** and **developing embryo**, particularly from desiccation (Angiosperm means "enclosed seed") so that reproductive success is more likely.

Endosperm is a store of nutrients for the developing plant embryo. It does not develop until fertilisation occurs so that food stores are not wasted, as they might be in gymnosperms.

Fruits are formed from ripened ovaries. Their specialised shapes, colours, smells and textures aid seed dispersal by wind, water and animals.

Large leaf surfaces are typical. Rapid growth and fruit production are very demanding on supplies of energy and raw material. Thus rates of photosynthesis must be high and small, needle-like leaves of gymnosperms such as pines would not be sufficient.

NB Greater leaf surfaces do bring the penalty of more loss of water by evaporation and diffusion through stomata.

Stomata with guard cells to regulate loss of water vapour and movements of oxygen and carbon dioxide between plant and atmosphere.

Waterproof cuticle to reduce water loss to atmosphere.

Xylem vessels permit much more efficient water conduction than the **tracheids** found in many other plants, since there are not end walls to impede movement.

Vascular system to transport water, ions and organic solutes.

Specialised supporting tissues since air is a medium of low density.

ANGIOSPERMS
show basic plant **adaptations to life on land**

Roots form beneficial associations with other organisms. For example, legumes form root nodules with nitrogen-fixing *Rhizobium* bacteria and many Angiosperms form ion-absorbing mycorrhizae with some fungi.

Extensive root systems for anchorage of erect shoot systems and absorption of water and ions.

Plant and animal cells

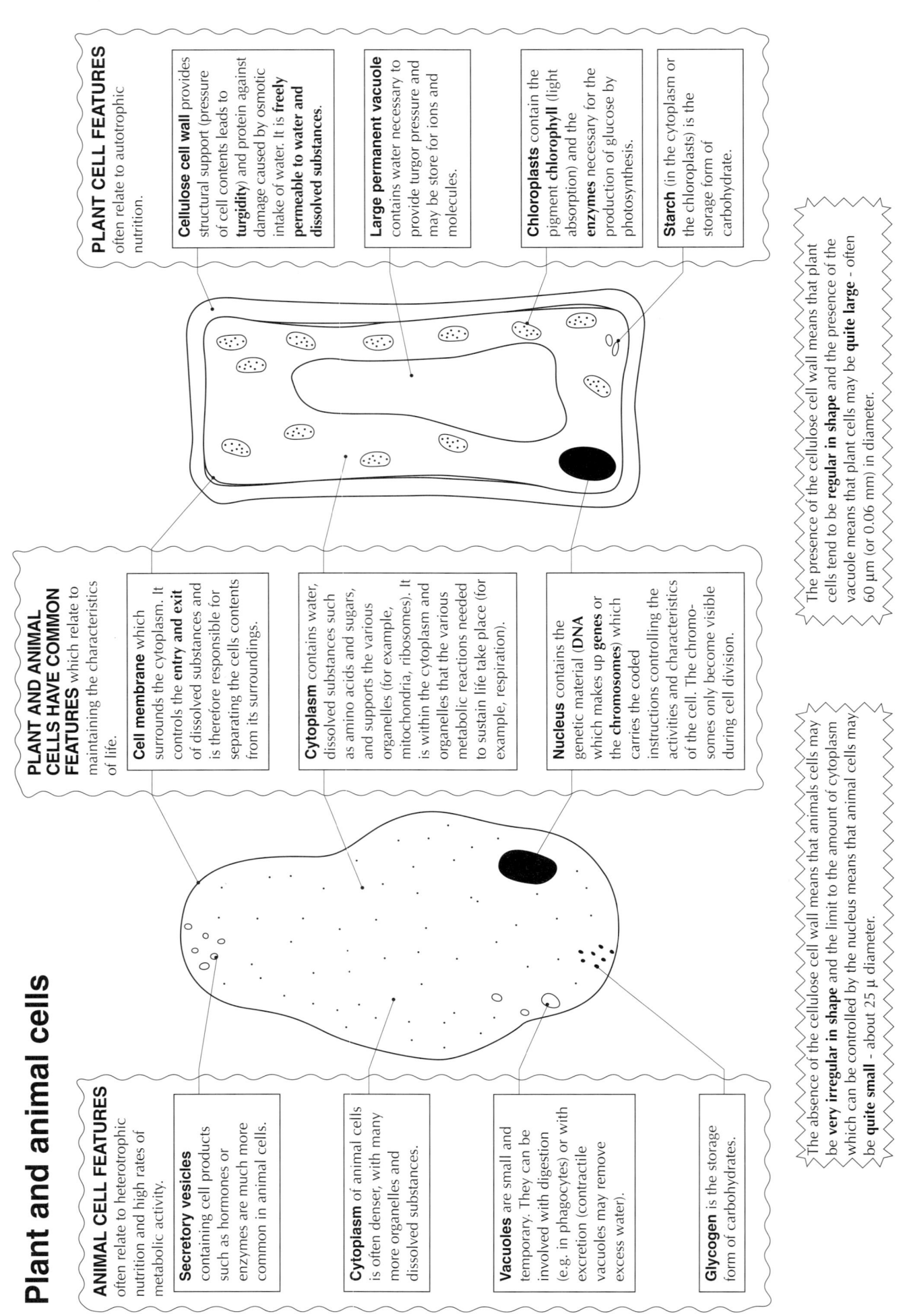

PLANT CELL FEATURES often relate to autotrophic nutrition.

Cellulose cell wall provides structural support (pressure of cell contents leads to **turgidity**) and protein against damage caused by osmotic intake of water. It is **freely permeable to water and dissolved substances**.

Large permanent vacuole contains water necessary to provide turgor pressure and may be store for ions and molecules.

Chloroplasts contain the pigment **chlorophyll** (light absorption) and the **enzymes** necessary for the production of glucose by photosynthesis.

Starch (in the cytoplasm or the chloroplasts) is the storage form of carbohydrate.

PLANT AND ANIMAL CELLS HAVE COMMON FEATURES which relate to maintaining the characteristics of life.

Cell membrane which surrounds the cytoplasm. It controls the **entry and exit** of dissolved substances and is therefore responsible for separating the cells contents from its surroundings.

Cytoplasm contains water, dissolved substances such as amino acids and sugars, and supports the various organelles (for example, mitochondria, ribosomes). It is within the cytoplasm and organelles that the various metabolic reactions needed to sustain life take place (for example, respiration).

Nucleus contains the genetic material (**DNA** which makes up **genes** or the **chromosomes**) which carries the coded instructions controlling the activities and characteristics of the cell. The chromosomes only become visible during cell division.

ANIMAL CELL FEATURES often relate to heterotrophic nutrition and high rates of metabolic activity.

Secretory vesicles containing cell products such as hormones or enzymes are much more common in animal cells.

Cytoplasm of animal cells is often denser, with many more organelles and dissolved substances.

Vacuoles are small and temporary. They can be involved with digestion (e.g., in phagocytes) or with excretion (contractile vacuoles may remove excess water).

Glycogen is the storage form of carbohydrates.

The presence of the cellulose cell wall means that plant cells tend to be **regular in shape** and the presence of the vacuole means that plant cells may be **quite large** - often 60 μm (or 0.06 mm) in diameter.

The absence of the cellulose cell wall means that animals cells may be **very irregular in shape** and the limit to the amount of cytoplasm which can be controlled by the nucleus means that animal cells may be **quite small** - about 25 μ diameter.

8 Plant and animal cells

Diffusion, osmosis and active transport

are processes by which molecules are moved. Diffusion and osmosis are passive, but active transport requires energy.

DIFFUSION:

the movement of ions or molecules down a concentration gradient i.e. from a region of higher concentration to one of lower concentration.

This is a physical process which depends on the energy possessed by the molecules, thus
- small molecules diffuse faster than large molecules
- diffusion speeds up as temperature increases.

For living cells the principle (the movement of molecules down a concentration gradient) is the same, but there is one problem → the cell is surrounded by a **cell membrane** which can restrict the free movement of the molecules

→ This is **a selectively permeable membrane**: the composition of the membrane (lipid and protein) allows some molecules to cross with ease, but others with difficulty or not at all. In this example the membrane is permeable to water ● but not to the larger sucrose molecule ⊘ – the simplest sort of selection is based on the **size** of the molecules.

Important examples are
- Oxygen from air sacs in the lung to blood, and from blood to cells
- Soluble foods from gut to blood
- Carbon dioxide from air to spaces inside leaf

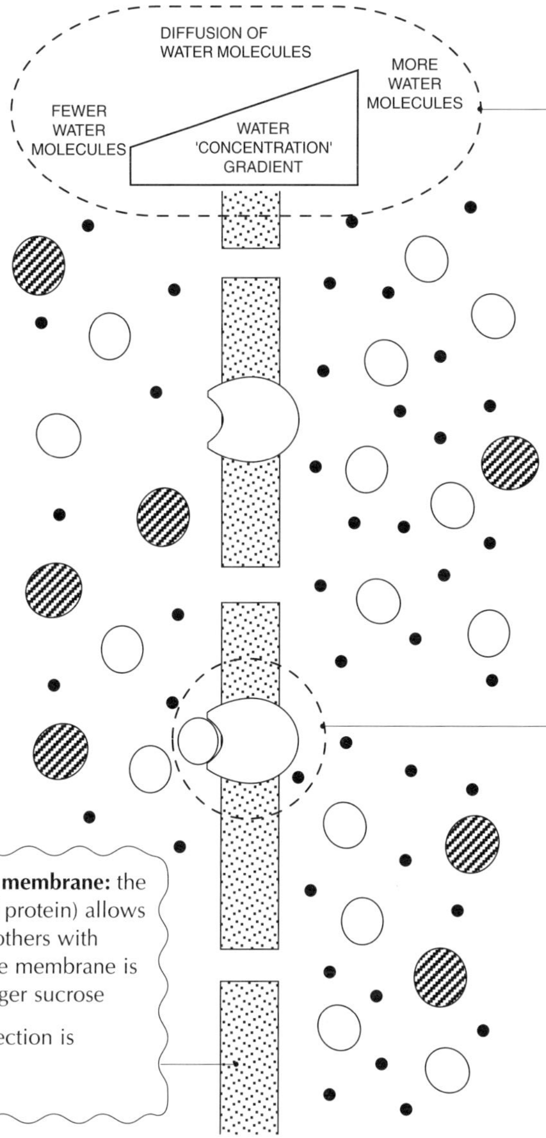

OSMOSIS IS THE DIFFUSION OF WATER

Water crosses membranes very freely and always tends to move, by diffusion, down the water 'concentration' gradient. The term 'concentration' can be confusing when used to describe water molecules, and is better replaced by the term 'potential'.

Thus osmosis is
- the movement of water
- across a selectively permeable membrane
- down a water potential gradient

Osmosis is responsible for water movement
- from tissue fluid to cells
- from soil water to root hairs
- from xylem to leaf mesophyll cells

ACTIVE TRANSPORT MAY MOVE MOLECULES AGAINST A CONCENTRATION GRADIENT

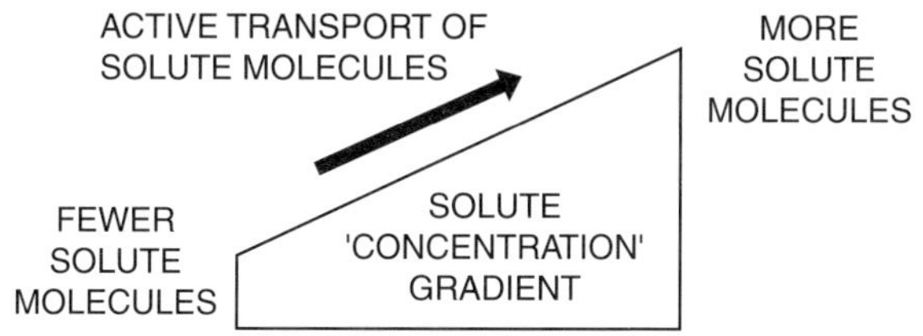

In this example there are more amino acid molecules on the right side of the membrane than on the left - to move any more from left to right will be 'uphill' - this active transport
- requires energy to 'drive' the molecules 'uphill' - this energy is supplied as ATP from respiration
- is affected by any factor which affects respiration, e.g. temperature and oxygen concentration
- is carried out by 'carrier proteins' in the membrane, which bind to the solute molecule, change shape, and carry the molecule across the membrane.

Important examples are
- uptake of mineral ions from soil by root hair cells
- movement of sodium ions to set up nerve impulses

Enzymes control biological process

and are widely exploited by humans.

ENZYMES ARE PROTEINS WHICH ACT AS CATALYSTS IN LIVING ORGANISMS

Substrate molecules fit exactly onto an ACTIVE SITE on the enzyme

Substrate molecules react together to form a **product** which leaves the active site

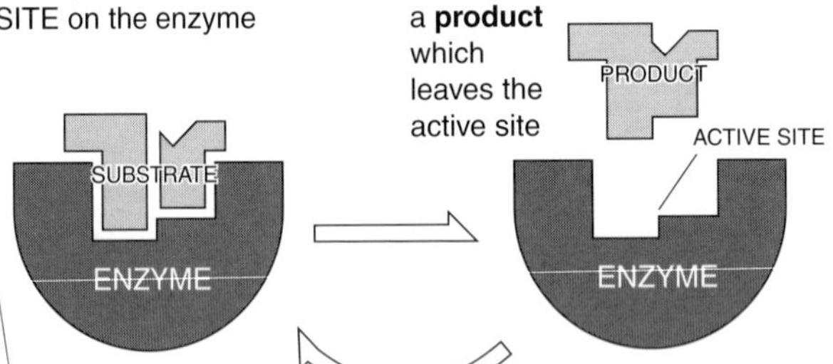

Enzyme molecule is now free to bind to more substrate molecules

The **shape** of the active site enables the enzyme to 'recognise' its substrate **in a very specific way.** Any factor which alters the enzyme's shape will affect its activity.

Inhibitors and activators: these are molecules which may
- **inhibit** by blocking the active site e.g. cyanide poisons by blocking enzymes in respiration
- **activate** by helping the active site to achieve its correct shape e.g. chloride ions in saliva activate the starch digesting enzyme salivary amylase

Enzymes may be intracellular or extracellular

these are both made **and** have their action inside cells ('intra' means 'inside')

e.g. photosynthetic enzymes inside chloroplasts respiratory enzymes inside mitochondria.

these are **made** inside cells but **have their action outside the cell** ('extra' means 'outside')

e.g. digestive systems in the human gut enzymes released by saprotrophic fungi and bacteria

THE SPECIFICITY AND CATALYTIC ACTIVITY OF ENZYMES MAKES THEM VERY USEFUL TO HUMANS

MEDICINE
- **Streptokinase** limits damage caused by heart attacks by dissolving blood clots
- **Urease** breaks down urea in dialysis fluid from kidney dialysis machines, allowing the fluid to be reused
- **Lactase** removes lactose from milk, thus making it safe for lactose-intolerant people

COMMERCIAL
- **Proteases** help to soften leather for the garment industry
- **Lipase** removes stains from clothing - component of 'biological' washing powders
- **Amylase** converts starch to sugars used in production of syrups, e.g. in fruit pies

PHARMACEUTICAL
- **Proteases** remove protein stains from false teeth
- **Catalase** removes hydrogen peroxide used to sterilise contact lenses

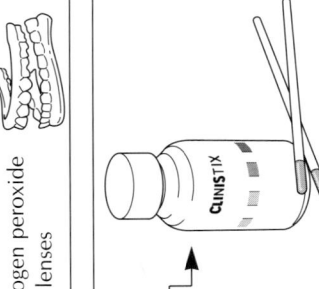

ANALYTICAL
- **Glucose oxidase** used in biosensors detects glucose levels in blood and urine
- **Carbonic anhydrase** detects levels of insecticides in water

GENETIC ENGINEERING
- **Restriction enzymes** are used to cut out specific genes, and to open up bacterial plasmids
- **Ligases** are used to 'stitch' human genes into bacterial plasmids

Other, related topics can be found on pages 11, 46.

TEMPERATURE: like all proteins, enzymes are made up of long, precisely folded chains of amino acids. This folding may be 'undone' by high temperature so that the enzyme may lose its active site - it is **denatured**.

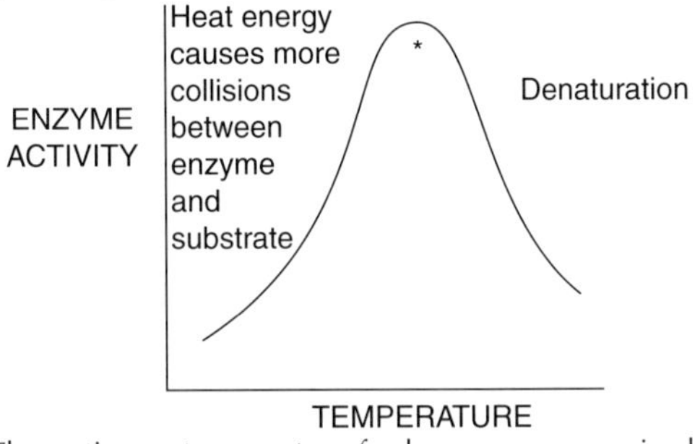

Heat energy causes more collisions between enzyme and substrate

Denaturation

ENZYME ACTIVITY

TEMPERATURE

* The optimum temperature for human enzymes is close to 37°C. For most plants it is lower.

pH: Is a measure of acidity or alkalinity, and is a mathematical method for expressing the concentration of H^+ ions in a solution.

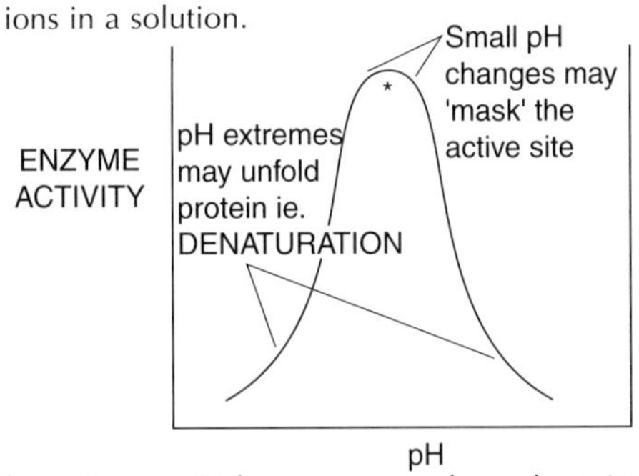

Small pH changes may 'mask' the active site

pH extremes may unfold protein ie. DENATURATION

ENZYME ACTIVITY

pH

* the optimum pH for an enzyme depends on its site of action e.g. enzymes in the stomach (where HCl is present) have an optimum about pH2 but intestinal enzymes (no HCl) have optimum pH about 7.5

An ideal human diet

contains fat, protein, carbohydrate, vitamins, minerals, water and fibre **in the correct proportions**.

An adequate diet provides sufficient **energy** for the performance of metabolic work, although the 'energy' could be in any form.

A balanced diet provides all dietary requirements **in the correct proportions**. Ideally this would be $\frac{1}{7}$ **fat**, $\frac{1}{7}$ **protein** and $\frac{5}{7}$ **carbohydrate**.

In conditions of **undernutrition** the first concern is usually provision of an **adequate diet**, but to avoid symptoms of **malnutrition** a **balanced diet** must be provided.

PROTEINS

Are **building blocks** for growth and repair of many body tissues (e.g. myosin in muscle, collagen in connective tissues), as **enzymes**, as **transport systems** (e.g. haemoglobin), as **hormones** (e.g. insulin) and as **antibodies**.

Common source: meat, fish, eggs and legumes/pulses. Must contain eight **essential amino acids** since humans are not able to synthesise them. (Animal sources generally contain more of the essential amino acids than vegetable sources). Digested in stomach and absorbed as **amino acids**.

Deficiency of protein causes poor growth - in extreme cases (in developing countries) may cause **marasmus** or **kwashiorkor**.

WATER

is required as a solvent, a transport medium, a substrate in digestive reactions and for lubrication (e.g. in tears). A human requires 2–3dm³ of water daily - most commonly from drinks and liquid foods.

MINERALS

have a range of **specific** roles, and absence may cause **deficiency diseases**

	Source	Deficiency causes
e.g. Calcium (Ca^{2+})	Dairy products	Poor growth of bones and teeth
Iron (Fe^{2+})	Red meat, spinach	Anaemia - poor oxygen transport in blood

They are usually ingested with other foods, but supplements may be necessary (e.g. iron tablets are sometimes needed following menstruation).

CARBOHYDRATES

Principally as an energy source. **Respiratory substrate** oxidised to release **energy** for active transport, synthesis of macromolecules, cell division and muscle contraction.

Common sources: rice, potatoes, wheat and other cereal grains i.e. as **starch**, and as refined sugar, **sucrose** in food sweetenings and preservatives.

Digested in mouth and small intestine and absorbed as **glucose**.

LIPIDS

An energy source (they are highly reduced and therefore can be oxidised to release **energy**). Also important in **cell membranes** and as a component of **steroid hormones**.

Common sources: Meat and animal foods are rich in **saturated fats** and **cholesterol**, plant sources such as sunflower and soya are rich in **unsaturated fats**.

Digested in the small intestine and absorbed as **fatty acids and glycerol**.

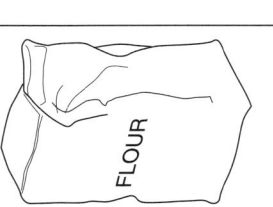

VITAMINS

have no common structure or function, but are essential in small amounts to use other dietary components efficiently. Their absence results in **deficiency diseases**.

		Source	Deficiency causes
e.g.	**vitamin C** (water soluble)	Citrus fruits	**Scurvy** - bleeding gums, slow healing of wounds
	vitamin D (fat soluble)	Cod liver oil, margarine	**Rickets** - misshapen and poorly growing bones

FIBRE

(originally known as **roughage**) is mainly cellulose from plant cell walls and is common in fresh vegetables and cereals. It **may** provide some energy but mainly serves to aid faeces formation and prevent constipation.

Ideal human diet 11

Malnutrition

Malnutrition in developing countries usually means **undernutrition** - the diet may be neither balanced nor adequate.

Vitamin A deficiency causes **night blindness** (poor production of visual pigment) and **xerophthalmia** (hardening and flaking of the cornea).

Iron deficiency causes **anaemia**. Iron is required for haemoglobin synthesis so insufficient dietary iron causes oxygen transport by red blood cells to be limited and energy release by aerobic respiration is reduced.

Vitamin C deficiency causes **scurvy**. The production of some structural tissues is inhibited so that teeth become loose, skin becomes flaky and wounds bleed freely. Very low levels of vitamin C may also make the immune system less efficient so that the body becomes more prone to infection.

Protein deficiency may cause **kwashiorkor** - mental and physical development is slowed down. The swollen abdomen which occurs in such cases is due to insufficient blood protein to be able to reclaim water from the tissues and to an enlarged liver. If there is also a shortage of energy foods **marasmus** may occur - this is a general wasting of all body tissues and the body becomes very thin and wrinkled.

Deficiency of calcium or vitamin D may cause poor development of bones - this is seen as **rickets** in children (characterised by 'bow legs') and **osteoporosis** in adults (often leading to easy fracturing and poor healing of bones).

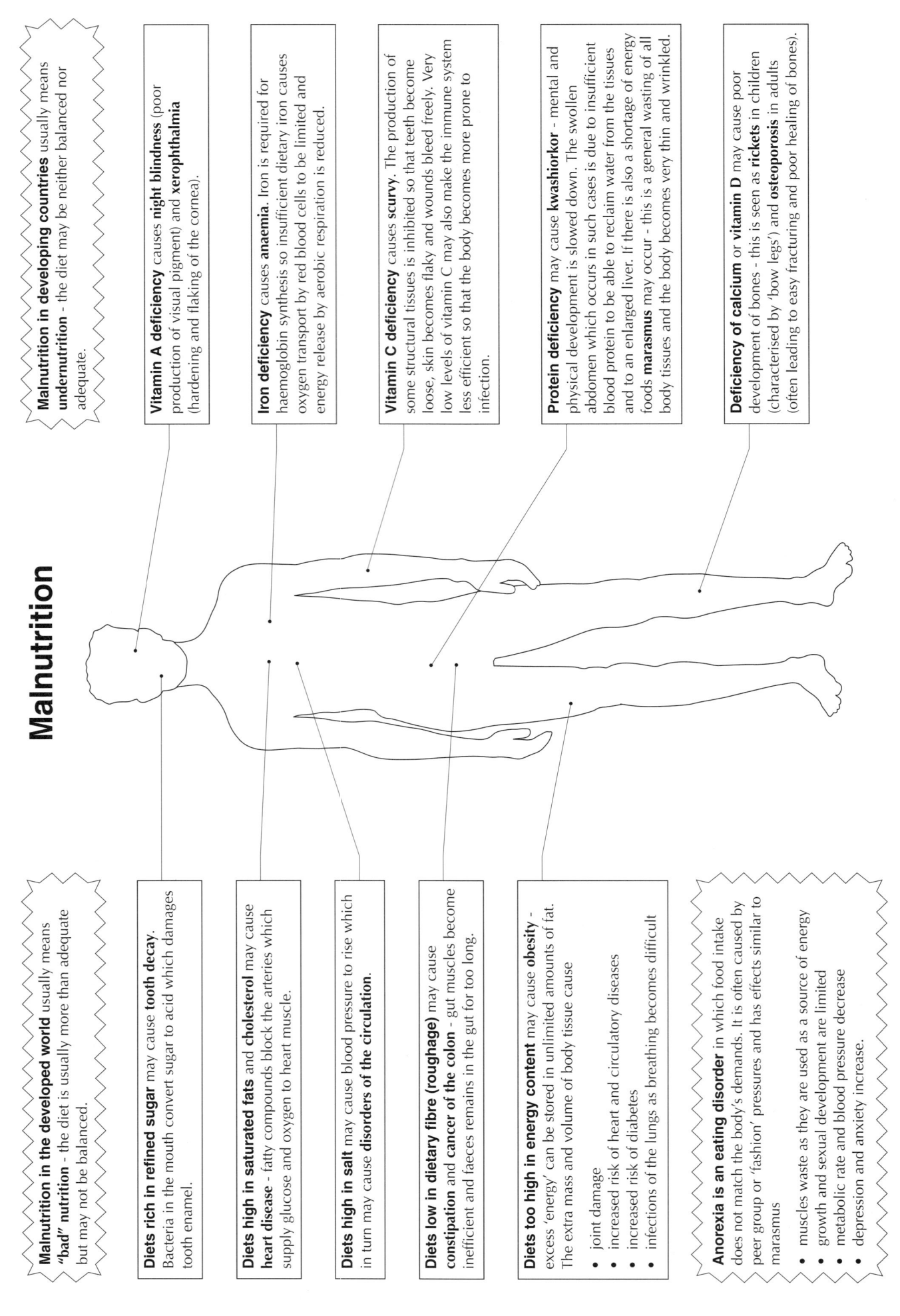

Malnutrition in the developed world usually means "bad" nutrition - the diet is usually more than adequate but may not be balanced.

Diets rich in refined sugar may cause **tooth decay**. Bacteria in the mouth convert sugar to acid which damages tooth enamel.

Diets high in saturated fats and **cholesterol** may cause **heart disease** - fatty compounds block the arteries which supply glucose and oxygen to heart muscle.

Diets high in salt may cause blood pressure to rise which in turn may cause **disorders of the circulation**.

Diets low in dietary fibre (roughage) may cause **constipation** and **cancer of the colon** - gut muscles become inefficient and faeces remains in the gut for too long.

Diets too high in energy content may cause **obesity** - excess 'energy' can be stored in unlimited amounts of fat. The extra mass and volume of body tissue cause
- joint damage
- increased risk of heart and circulatory diseases
- increased risk of diabetes
- infections of the lungs as breathing becomes difficult

Anorexia is an eating disorder in which food intake does not match the body's demands. It is often caused by peer group or 'fashion' pressures and has effects similar to marasmus
- muscles waste as they are used as a source of energy
- growth and sexual development are limited
- metabolic rate and blood pressure decrease
- depression and anxiety increase.

Digestion, absorption and processing of foods

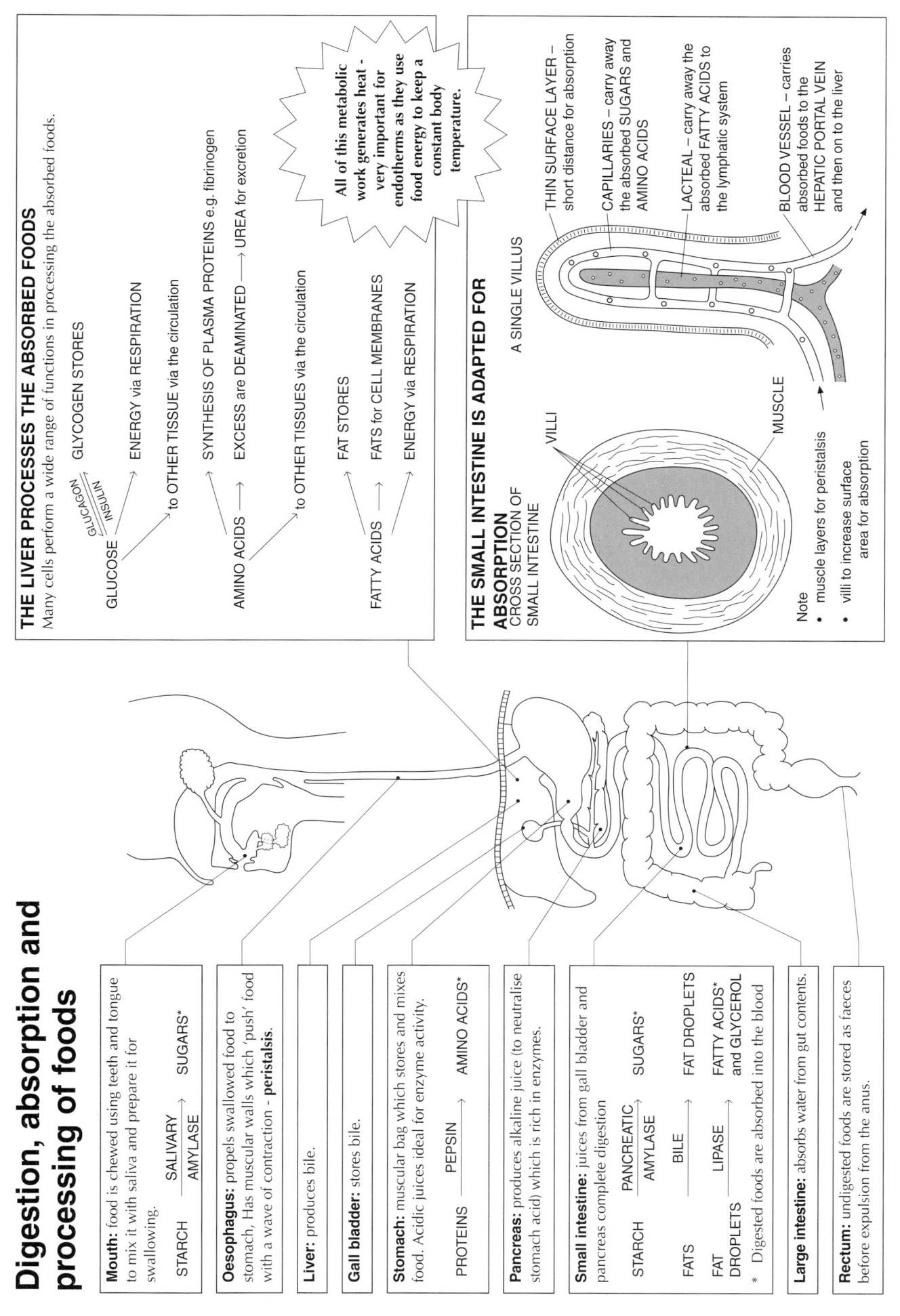

Mouth: food is chewed using teeth and tongue to mix it with saliva and prepare it for swallowing.

STARCH $\xrightarrow{\text{SALIVARY AMYLASE}}$ SUGARS*

Oesophagus: propels swallowed food to stomach, Has muscular walls which 'push' food with a wave of contraction – **peristalsis**.

Liver: produces bile.

Gall bladder: stores bile.

Stomach: muscular bag which stores and mixes food. Acidic juices ideal for enzyme activity.

PROTEINS $\xrightarrow{\text{PEPSIN}}$ AMINO ACIDS*

Pancreas: produces alkaline juice (to neutralise stomach acid) which is rich in enzymes.

Small intestine: juices from gall bladder and pancreas complete digestion

STARCH $\xrightarrow{\text{PANCREATIC AMYLASE}}$ SUGARS*

FATS $\xrightarrow{\text{BILE}}$ FAT DROPLETS

FAT DROPLETS $\xrightarrow{\text{LIPASE}}$ FATTY ACIDS* and GLYCEROL

* Digested foods are absorbed into the blood

Large intestine: absorbs water from gut contents.

Rectum: undigested foods are stored as faeces before expulsion from the anus.

THE LIVER PROCESSES THE ABSORBED FOODS
Many cells perform a wide range of functions in processing the absorbed foods.

GLUCOSE $\underset{\text{INSULIN}}{\overset{\text{GLUCAGON}}{\rightleftarrows}}$ GLYCOGEN STORES

GLUCOSE → ENERGY via RESPIRATION

→ to OTHER TISSUE via the circulation

→ SYNTHESIS OF PLASMA PROTEINS e.g. fibrinogen

AMINO ACIDS → EXCESS are DEAMINATED → UREA for excretion

→ to OTHER TISSUES via the circulation

FATTY ACIDS → FAT STORES

→ FATS for CELL MEMBRANES

→ ENERGY via RESPIRATION

All of this metabolic work generates heat – very important for endotherms as they use food energy to keep a constant body temperature.

THE SMALL INTESTINE IS ADAPTED FOR ABSORPTION

CROSS SECTION OF SMALL INTESTINE

A SINGLE VILLUS

THIN SURFACE LAYER – short distance for absorption

CAPILLARIES – carry away the absorbed SUGARS and AMINO ACIDS

LACTEAL – carry away the absorbed FATTY ACIDS to the lymphatic system

BLOOD VESSEL – carries absorbed foods to the HEPATIC PORTAL VEIN and then on to the liver

VILLI

MUSCLE

Note
- muscle layers for peristalsis
- villi to increase surface area for absorption

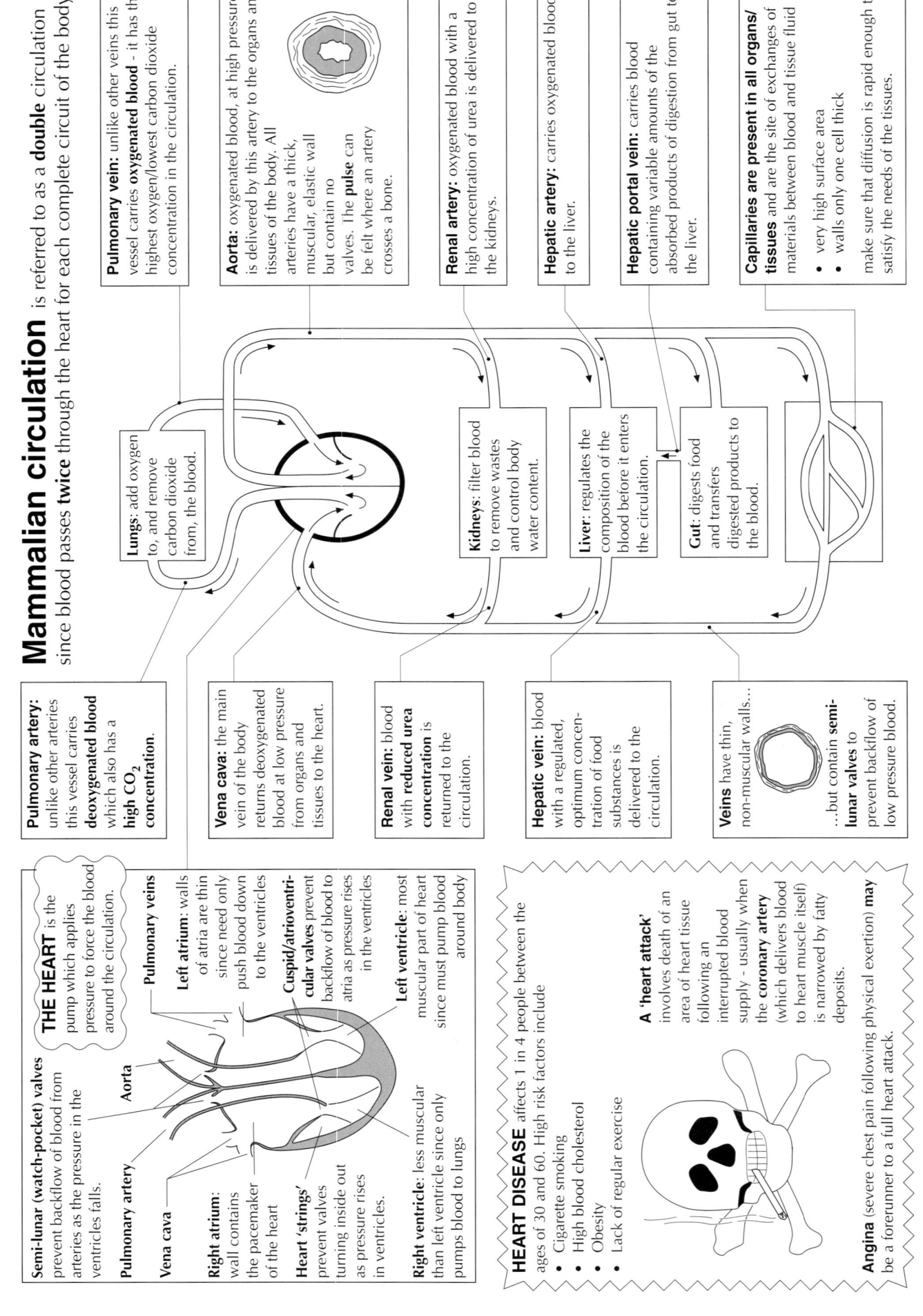

Blood structure and function

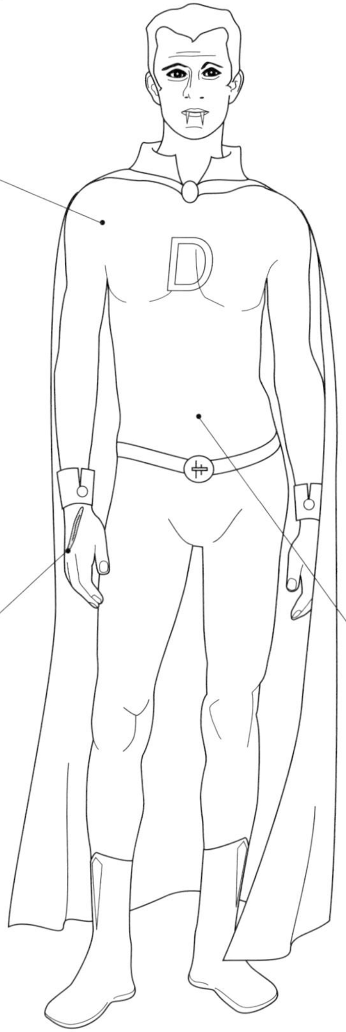

COMPONENTS OF BLOOD
if a blood sample is allowed to settle, or spun for a few minutes in a centrifuge, it separates into two distinct layers

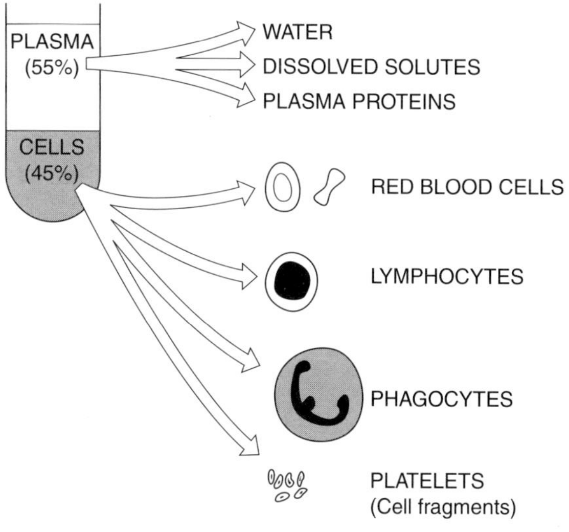

- PLASMA (55%)
  - WATER
  - DISSOLVED SOLUTES
  - PLASMA PROTEINS
- CELLS (45%)
  - RED BLOOD CELLS
  - LYMPHOCYTES
  - PHAGOCYTES
  - PLATELETS (Cell fragments)

TRANSPORT FUNCTIONS

Soluble products of digestion/absorption (such as glucose, amino acids, fatty acids, vitamins and minerals) from the gut to the liver and then to the general circulation.

Waste products of metabolism (such as urea, creatinine and lactate) from sites of production to sites of removal, such as the liver and kidney.

Respiratory gases (oxygen and carbon dioxide) from their sites of uptake or production to their site of use or removal. Oxygen transport is more closely associated with red blood cells, and carbon dioxide transport with the plasma.

Hormones (such as insulin) from their sites of production in the glands to the target organs where they exert their effects.

REGULATORY FUNCTIONS - HOMEOSTASIS

Blood solutes affect the water potential of the blood, and thus the water potential gradient between the blood and the tissue fluid. The size of this water potential gradient is largely due to sodium ions and plasma proteins and **regulates the movement of water** between blood and tissues.

Water plays a part in the **distribution of heat** between heat producing areas such as the liver and areas of heat loss such as the skin.

Too much blood alcohol can cause water to leave brain cells causing the pain and sensation of thirst called a **hangover!**

Blood also helps to maintain **optimum pH** in the tissues.

PROTECTIVE FUNCTIONS

Platelets, plasma proteins (e.g. fibrinogen) and many other plasma factors (e.g. Ca^{2+} ions) protect against **blood loss** and the **entry of pathogens** by the clotting mechanism.

White blood cells protect against disease-causing organisms:
phagocytes engulf them
lymphocytes produce and secrete specific antibodies against them.

Dracula may have consumed blood because he suffered from **porphyria**, a condition characterised by an inability to synthesise some compounds including the haem group of haemoglobin, a marked sensitivity to daylight and abnormal development of the teeth!

Disease may have a number of causes

SOME DISEASES ARE NOT INFECTIOUS.
They are **not** caused by pathogens. They may be

Degenerative: organs and tissues may work less well as they age. This is thought to be due to changes to body chemicals caused by **free radicals** such as the peroxide ion e.g. heart attacks, cataracts, hardening of the arteries.

Deficiency: poor diet may deprive the body of some essential substance e.g. scurvy (lack of vitamin C).

Allergy: sensitivity to some environmental antigen e.g. hayfever (pollen is the antigen).

Environmental: some factor in the environment may trigger a dangerous or abnormal bodily reaction e.g. overexposure to ultraviolet radiation may cause abnormal cell division leading to skin cancer.

Inherited/metabolic: some failure in the body's normal set of chemical reactions e.g. sickle cell anaemia (abnormal haemoglobin) cystic fibrosis (production of thick mucus) diabetes (failure to produce enough insulin) These conditions are due to alterations in the genes.

Psychological/mental: some changes in the working of the brain may lead to abnormal behaviour e.g. schizophrenia, depression.

Self-induced: some abuse of the body may affect its function e.g. lung cancer (cigarette smoking), cirrhosis of the liver (alcohol abuse).

Cure (i.e. treat the patient with the disease)
- **Antibiotics** to destroy bacteria
- **Passive immunity** to provide ready-made antibodies
- **Specific drugs** to kill abnormal cells
- **Pain relievers** to limit discomfort/speed recovery.
- **Radiotherapy** to prevent division of abnormal cells.

Typical symptoms of disease include
- **sweating/fever** due to resetting of body's thermostat
- **vomiting/diarrhoea** due to attempt to 'clear' gut of irritants
- **pain** due to release of toxins by pathogens.

During **disease** the body's activities deviate from their normal levels by an amount which is more than can be counteracted by the usual homeostatic mechanisms.

INFECTIOUS DISEASES ARE CAUSED BY PATHOGENS - these may spread in a number of ways:

in droplets in the air e.g. Influenza virus

in contaminated food e.g. *Salmonella* bacteria

in infected water e.g. *Cholera* bacteria

by direct contact (contagious) e.g. Athlete's foot-fungus

via body fluids e.g. Hepatitis B virus

by animal vectors e.g. *Plasmodium* protozoan (causes malaria) via *Anopheles* mosquito

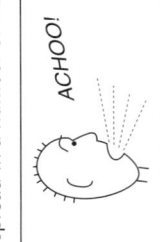

INFECTIOUS DISEASE MAY BE CONTROLLED:

Prevention (i.e. do not contract the disease)
- Sanitation to protect water supplies
- Effective farming, e.g. of chickens, to prevent food poisoning
- Hygiene to prevent spread of contagious diseases. Use of **antiseptics** to kill pathogens before they enter the body.
- Destruction of vectors such as mosquitoes with insecticides
- Vaccination to 'prime' the immune system.

Food preservation

may prevent disease

Saprotrophs e.g. fungi and bacteria are responsible for the decomposition of food. This causes two problems:

- as the food decays **less is available for humans**, either because the food has been 'consumed' or because its texture or appearance become less attractive.
- the saprotroph may release compounds onto the food – these may be **toxins** (poisons) which can cause discomfort or illness.

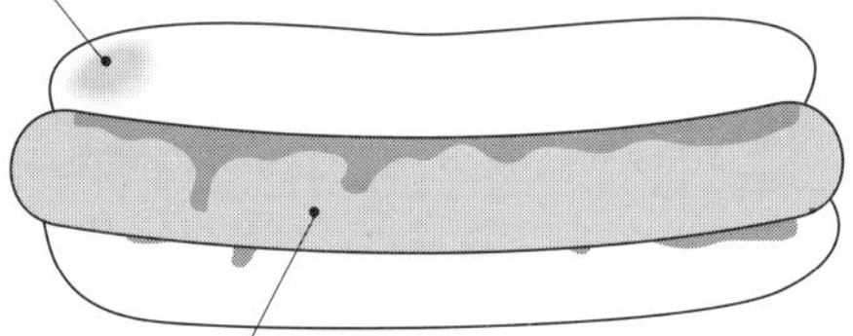

Milk pasteurisation is a valuable technique since
- it destroys most bacterial cells
- it does not alter the flavour of the milk.

The **flash** process involves
- heating food rapidly to 72°C (**not boiling**)
- hold at this temperature for 15 seconds
- cool rapidly (and keep **refrigerated**).

TECHNIQUES OF FOOD PRESERVATION

Sterilisation
- Food heated to high temperature, then sealed in an airtight container
- destroys bacterial cells, but not their spores
- may alter the taste of the food.

Refrigeration
- keep food between 0–4°C
- does not **kill** bacteria, but slows their reproduction – thus the food is only preserved for 48–72 hours.

Irradiation
- food is exposed to γ-radiation
- microorganisms are destroyed, but enzymes are not. Thus irradiated food is not affected in texture or in ripening time, and this makes this technique attractive to customers.

Irradiated food does not contain microorganisms but as in other preservation techniques, **toxins are not destroyed** - thus irradiated food may still be dangerous.

Chemical methods
- use preservative substances to kill bacteria or slow down their reproduction
- examples include sulphur dioxide and nitrate compounds. Such substances are **additives** and can be identified by having an 'E number' e.g. E270.

A water supply is necessary for **drinking, washing** and **preparation of food**.

WATER MUST BE STERILISED

- Important water-borne diseases include:
- **Typhoid** (transmitted in faeces of infected people
- **Cholera** (water and shellfish may be contaminated)
- **Dysentery** (can be caused by bacteria or by an amoeba which is transmitted in infected faeces)
- Water is purified by **sewage treatment** and by the use of **chlorine**, which directly kills the infecting organisms.

***Salmonella* can cause food poisoning and can occur in undercooked meat or infected eggs.**

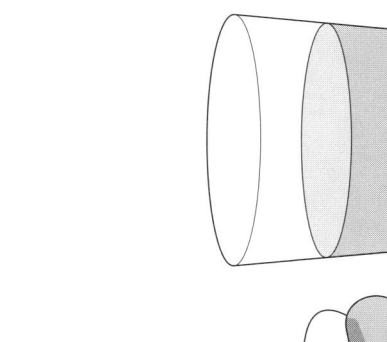

Intensive rearing of chickens makes it easy for *Salmonella* bacteria to spread from chicken gut → faeces → floor → chicken carcasses.

68°C cooking to this temperature kills *Salmonella* in meat.

Treatment: antibiotics are **not** very effective because they do not easily reach the gut lining.

Replace body fluids with a mixture of water, glucose and salts.

Salmonella in gut multiplies under ideal conditions of temperature/food availability and releases a **toxin**. This inflames the gut lining → fever, pain, vomiting and diarrhoea → dehydration, and even death.

Food preservation 17

Natural defence systems of the body

prevent infection and disease.

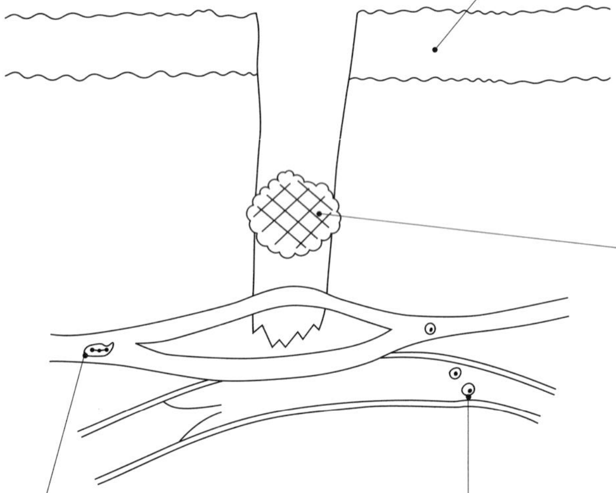

SKIN IS THE FIRST BARRIER:
the outer layer of the skin, the **epidermis**, is waxy and impermeable to water and to pathogens (although many microorganisms can live on its surface). Where there are natural 'gaps' in the skin there may be protective secretions to prevent entry of pathogens, e.g.

Orifice	Function	Protected by
Mouth	Entry of food	Hydrochloric acid in stomach
Eyes	Entry of light	Lysozyme in tears
Ears	Entry of sound	Bacteriocidal ('bacteria-killing') wax

BLOOD CLOTTING 'PLUGS' WOUNDS
is a natural defence, largely due to **blood proteins** and **platelets**. It is able to block any unnatural gaps in the skin.
- prevents excessive blood loss
- prevents entry of pathogens

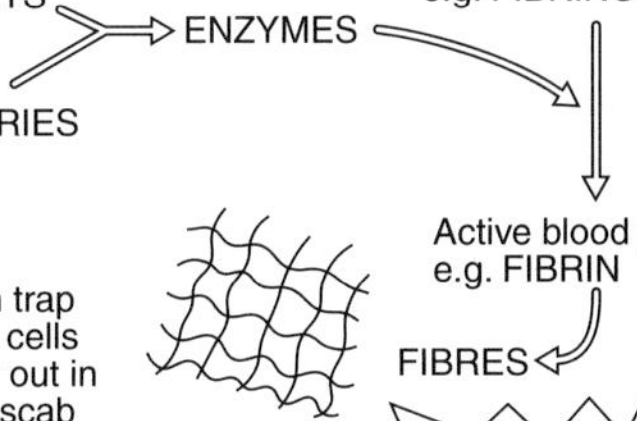

DAMAGED PLATELETS → ENZYMES

TORN CAPILLARIES

Inactive blood proteins e.g. FIBRINOGEN

If CALCIUM IONS present

Active blood proteins e.g. FIBRIN → FIBRES

Mesh can trap red blood cells which dry out in form of a scab

Absence of any of these blood proteins causes inefficient clotting → severe bleeding: **haemophilia**.

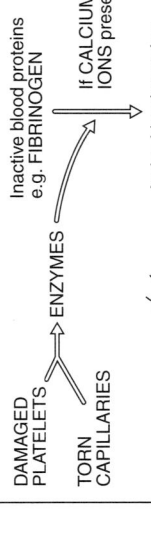

Surface protein identifies pathogen as 'non-human' cells

LYMPHOCYTES PRODUCE PROTECTIVE ANTIBODIES

Lymphocytes are white blood cells which are found both in the blood and in lymph nodes (swellings in the lymphatic system). They are stimulated by the presence of pathogens to manufacture and release special proteins called **antibodies** which can recognise, bind to, and help to destroy pathogens.

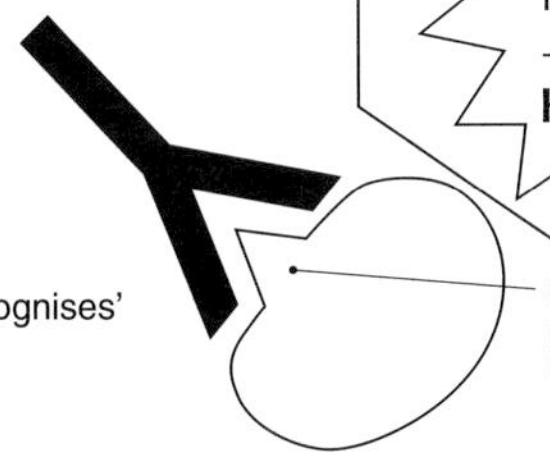

This end of antibody acts as a 'signal' to phagocytes to remove this pathogen

Forked end of antibody 'recognises' surface protein of pathogen

PATHOGENS CAUSE DISEASE.
Many organisms may colonise the body of a human - the body is **warm, moist** and a **good food source**. These organisms may compete with human cells for nutrients or may produce by-products which are poisonous to human cells. This will affect the normal function of the cells i.e. it will cause 'disease'. There are three major groups of pathogens:

Type of pathogen	Disease	Symptoms
VIRUS — Protein coat, Nucleic acid	Influenza	Fever (raised body temperature). Aching joints. Breathing problems. **Control by pain relief/rest/drinking fluids.**
BACTERIUM — Slime coat, Cell wall, 'Naked' DNA (not in chromosome)	Gonorrhoea	Painful to urinate - yellow discharge from penis or vagina. In the long-term may cause blockage of sperm ducts/oviducts leading to sterility. **Control with antibiotics.**
FUNGUS — DNA in nucleus, Hypha secretes enzymes into food	Athlete's foot	Irritation to moist areas of skin (e.g. between toes). 'Cracked' skin may become infected. **Control with fungicide/drying powders.**

PHAGOCYTES DESTROY PATHOGENS BY INGESTING THEM

Phagocytes are large white blood cells. They are 'attracted to' wounds or sites of infection by chemical messages. They leave the blood vessels and destroy any pathogens which they recognise.

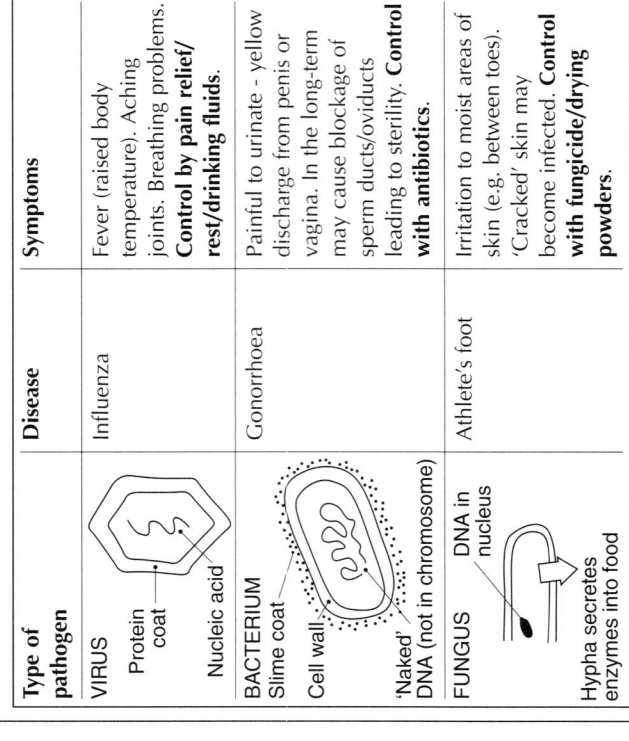

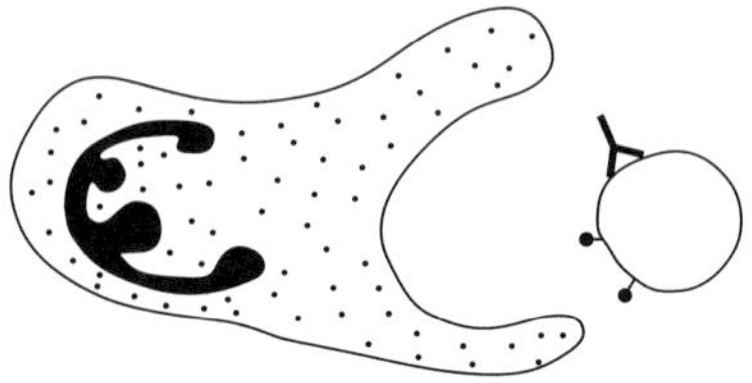

A pathogen is recognised by its surface proteins or by antibodies which 'label' it as dangerous.

Pathogens are destroyed by digestive enzymes secreted into 'food sac'.

Antibodies are specific protein molecules.

THE IMMUNE RESPONSE MAY BE ENHANCED.
Immunity is the body's use of antibodies to combat invasion by pathogens.

Immunity may be

ACTIVE: the individual is provoked to make his or her own antibodies

- **NATURAL** — Pathogen infects individual — Individual contracts disease but survives, makes antibodies and is now immune to further infection by the same pathogen e.g. immunity develops to different strains of the common cold.
- **ARTIFICIAL** — Weakened pathogen (vaccine) — Injection of vaccine does not cause disease but lymphocytes do produce antibodies – individual now immune to this pathogen e.g. vaccination against rubella virus (causes German measles) in babies.

PASSIVE: the individual is protected by a supply of pre-formed antibodies

- **NATURAL** — Mother's antibodies can cross the placenta and are delivered in breast milk: newborn child will be temporarily immune to pathogens for which mother produced antibodies.
- **ARTIFICIAL** — Antibodies collected from blood of laboratory animal e.g. horse or rabbit. INJECTION — Adult is now immune to disease which is too fast-acting for own immune system to deal with e.g. injection of anti-tetanus vaccine following a deep, dirty cut or wound. NB This offers only a TEMPORARY immunity.

VACCINE PRODUCTION
Use either
- Dead pathogens e.g. Whooping cough vaccine.
- Weakened pathogens e.g. Oral polio vaccine.
- Genetically engineered fragments – the proteins from the pathogens surface which are recognised by lymphocytes e.g. Hepatitis B viral coat protein

MONOCLONAL ANTIBODIES
are specific protein molecules with important medical and industrial uses

Injection of specific antigen into mouse provokes mouse immune system.

Specific lymphocytes isolated from mouse spleen. These cells produce antibodies but do not divide in tissue culture.

Tumour cells can divide rapidly in tissue culture but cannot produce antibodies.

FUSE TOGETHER

Hybridomas: these cells can produce specific antibodies **and** can divide rapidly in tissue culture.

SELECTION

Hybridoma producing only a single kind of antibody.

MONOCLONAL ANTIBODIES have many uses

- Pregnancy testing: the antibody detects female sex hormones in urine.
- Drug targetting: drugs can be 'attached to' a monoclonal antibody and targetted at a cell which the antibody can recognise.
- Protein purification: a useful protein can be extracted from a mixture since a monoclonal antibody can be made which binds only to the required protein

Abuse of drugs

MANY DRUGS AFFECT BRAIN FUNCTION

Barbiturates (taken as sleeping tablets) can depress breathing centre: **death**.

L.S.D. (a hallucinogen): affects ability to 'sort out' messages leading to faulty 'connections': **hallucinations**

Heroin: causes euphoria (a 'high') by depressing input of 'pain/discomfort' messages: **poor appetite, loss of awareness of surroundings**.

Caffeine (taken as a stimulant) increases sensitivity of nerve endings: **hyperactivity**.

OVER CONSUMPTION OF ALCOHOL has many harmful effects, both long- and short-term.

- inhibition of nerve impulse conduction: **slowed reflexes, dangerous driving, death**.
- depression of visual centre of brain: **poor co-ordination, accidents**.
- depression of breathing centre of brain: **death**.
- dilation of blood vessels of skin: **heat loss, hypothermia, death**.
- depression of appetite: **malnutrition**.
- irritation of lining of stomach: **vomiting - inhalation of vomit may cause death**.
- conversion to harmful product in liver causes hardening of lining of liver passages: **cirrhosis** - eventual liver failure: **death**.
- changes in blood pressure: **reduction in sexual performance**.

OVERUSE OF PAINKILLERS MAY BE DANGEROUS.

- excess paracetamol is converted to a poison by the liver, causing liver failure: **death**.
- excess aspirin reduces blood clotting ability: **bleeding of stomach lining**.

SOLVENT ABUSE causes both short- and long-term damage.

- damages membranes of brain cells: **loss of memory, irrational behaviour**.
- damages membranes which line the nose: **more infections of breathing passages**.
- very high doses of, e.g. butane, cause constriction of airways: **death by asphyxiation**.

SMOKING OF TOBACCO has many harmful effects.

- inhibits cilia which normally sweep a stream of mucus up the bronchi to trap and remove dust, pollen and pathogens. Lining of bronchi become irritated and infected: **bronchitis**.
- carcinogens provoke uncontrolled cell division in tissues of airways - tumours develop, block airways and invade adjacent tissues: **lung cancer**.
- changes function of white blood cells in lungs which become overactive and 'digest' lung tissue: **emphysema**.
- carbon monoxide irreversibly binds to haemoglobin and reduces oxygen-carrying ability of blood - heart 'pumps' harder to compensate: **heart failure**.
- increases deposition of fats in the arteries thus reducing circulation - tissues die: **limb amputations**.
- residues to be excreted are stored as part of urine in the bladder. Cells divide uncontrollably: **bladder cancer**.

OVERUSE OF ANTIBIOTICS may clear the harmless bacteria and fungi from the gut, allowing possibly dangerous organisms to colonise the now-vacant 'habitat': **severe gut infections, diarrhoea**.

ABUSE OF STEROID HORMONES may raise blood levels of these compounds so that they feedback inhibit natural production of sex hormones: **sterility, loss of sexual desire**.

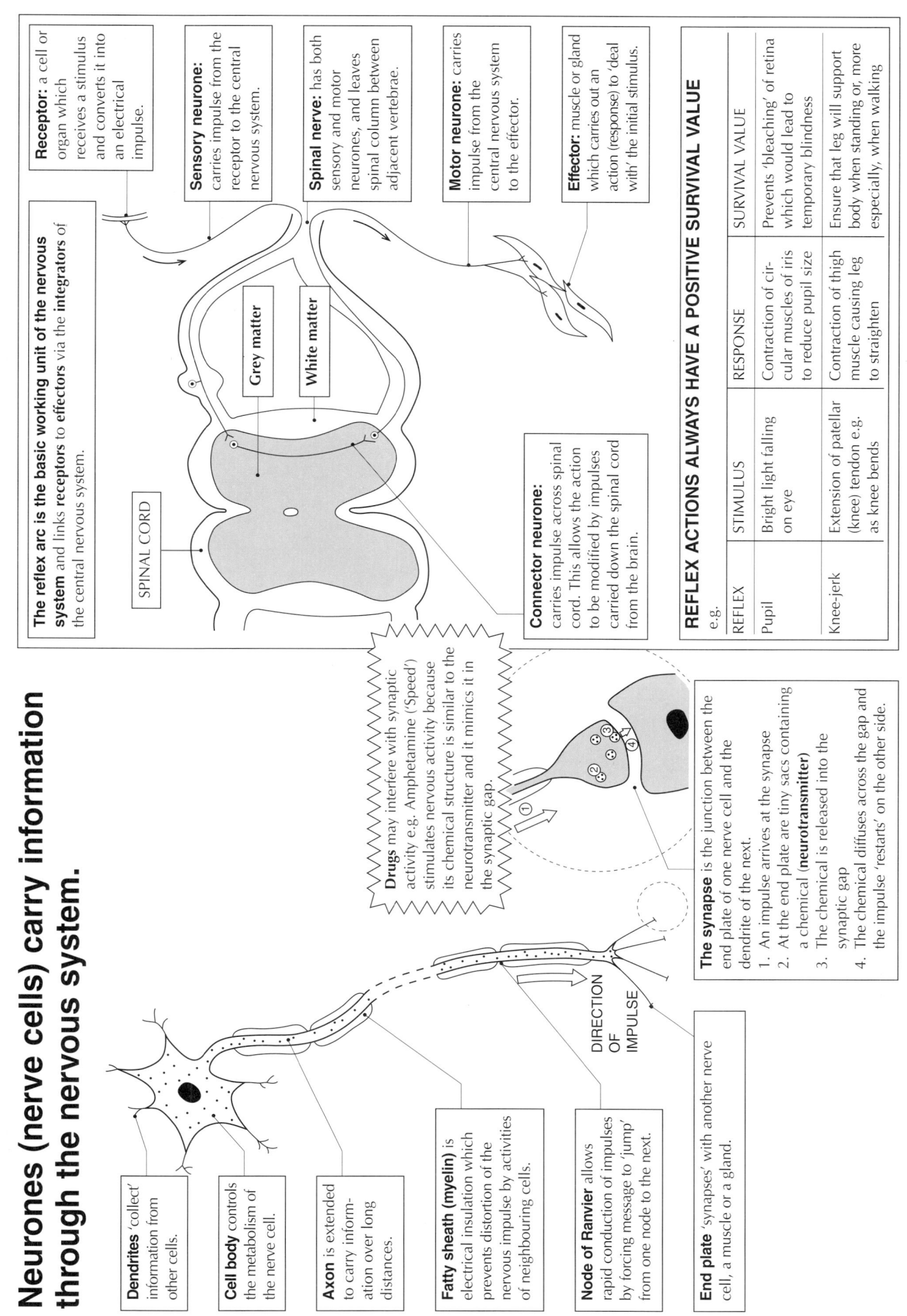

The endocrine system

produces and secretes chemical messengers called **hormones**.

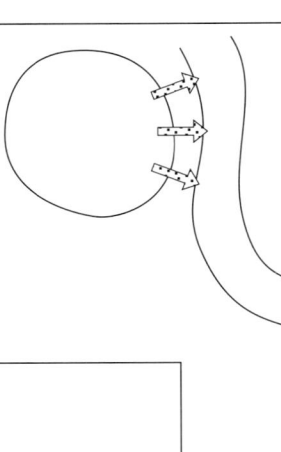

1. Stimulus affects endocrine gland so that it releases a hormone (chemical messenger).

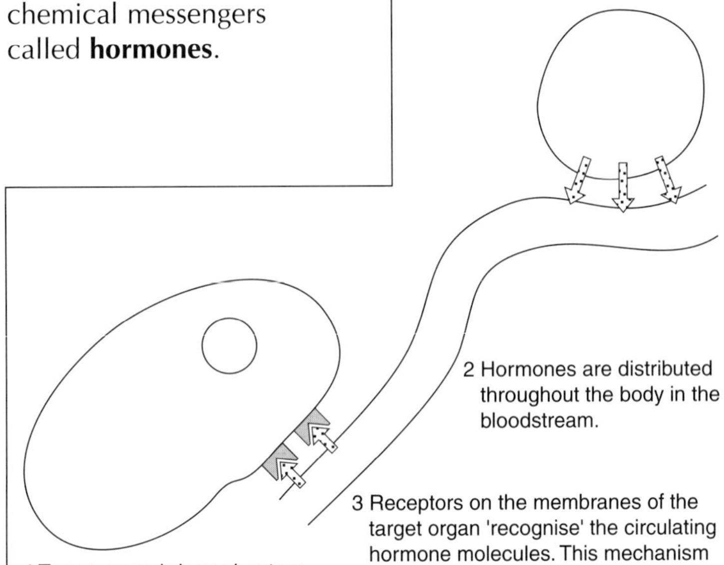

2. Hormones are distributed throughout the body in the bloodstream.

3. Receptors on the membranes of the target organ 'recognise' the circulating hormone molecules. This mechanism ensures that only the specific target organ can respond to the hormone.

4. Target organ brings about an appropriate response to deal with the original stimulus.

NERVOUS AND CHEMICAL CO-ORDINATION COMPARED

	NERVOUS	CHEMICAL (ENDOCRINE)
NATURE OF MESSAGE	Electrochemical impulses	Chemical compounds (hormones)
ROUTE OF TRANSMISSION	Specific nerve cells	General blood system
TYPE OF EFFECTS	Rapid, but usually short-term e.g. blinking	Usually slower, but generally longer-lasting e.g. growth

Note that the hormone **adrenaline** has effects which are rapid but short-term, and therefore partially mimics some effects of the nervous system.

SOME IMPORTANT APPLICATIONS OF HORMONES

- human growth hormone may be used to treat children who are growing very slowly
- insulin is used to control diabetes
- oestrogen and progesterone may be combined in the contraceptive pills to inhibit ovulation
- oestrogen may be used to promote milk production in cattle.

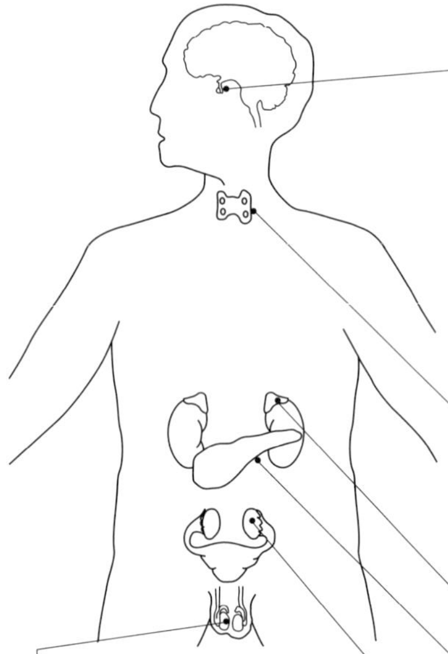

Adrenal glands secrete **adrenaline** which is called the **fight** or **flight hormone** because it prepares the body for action.

- more glycogen is converted to glucose
- deeper, more rapid breathing
- faster heartbeat
- diversion of blood from gut to muscles.

The above all help to provide more glucose and more oxygen for the working muscles.

There may also be
- dilation of the pupils
- raising of the hair.

Pituitary gland secretes
- a number of hormones (trophins) which influence other endocrine organs
- human growth hormone
- anti diuretic hormone.

Thyroid gland secretes thyroxine which controls metabolic rate and is important in regulation of body temperature.

Pancreas secretes insulin (reduces blood glucose) and glucagon (raises blood glucose concentration).

Testes secrete testosterone, which
- controls the production and development of sperm
- regulates the development of the male secondary sexual characteristics
i.e. enlargement of sex organs
growth of facial hair
muscle enlargement
deepening of voice
sexual behaviour

Artificial steroids increase muscle bulk but inhibit natural testosterone production which can lead to sterility.

Ovaries secrete the female sex hormones.

Oestrogen
- controls the development of the female secondary sexual characteristics, including the pattern of laying down of fat which gives the feminine body shape
- controls the repair of the lining of the uterus
- helps to regulate the development of further follicles.

Progesterone
- prepares the lining of the uterus for implantation by increasing the thickness of the lining and number of blood vessels
- helps to inhibit the growth of further follicles.

The eye is a sense organ –

different tissues work together to perform one function...

... **retina** contains light sensitive cells

... **cornea** and **lens** focus the light onto the retina

... **sclera**, **choroid** and **iris** protect and support the eye.

Accommodation: is the process of producing a finely-focused image on the retina. It is carried out by the action of the **ciliary muscles** on the **lens**.

Close object:
light must be **greatly refracted** (bent)
ciliary muscles **contract**, pull eyeball inwards
ligaments **relax**
lens becomes **short and fat**.

Distant object:
light can be **less refracted**
ciliary muscles **relax**, eyeball becomes spherical
ligaments **tauten**
lens is pulled **long and thin**.

Rods: provide **black-and-white** images but, because several may be 'wired' to a single sensory neurone in the optic nerve, they provide **great sensitivity** at low light intensity (night vision), but images lack detail.

Layer of pigment: prevents internal reflection which might lead to 'multiple/blurred' images.

Cones: provide very **detailed** images, in **colour** (there are three types, sensitive to red, green and blue light) but **only under high light intensity** (their connections to the optic nerve make them rather insensitive).

Choroid: a darkly coloured layer which (a) limits internal reflections and (b) contains blood vessels which help to nourish the cells of the retina.

Sclera: the tough outer coat which protects the eye against damage and provides attachment for the muscles which move the eye in its socket.

Retina: contains the light-sensitive cells, the rods and cones. Light arriving at this layer will produce an **inverted, smaller image**.

Yellow spot (Fovea): has the greatest density of cones and thus offers **maximum sharpness** but only works at full efficiency in **bright light**.

Optic nerve: composed of sensory neurones which carry nervous impulses to the **visual centre** at the rear of the brain.

Blind spot: corresponds to the exit point for the optic nerve. There are no light-sensitive cells here so that light falling on this region cannot be detected.

Vitreous humour: a jelly-like substance which helps to keep the shape of the eyeball, supports the lens and keeps the retina in place at the rear of the eye.

Suspensory ligament

Ciliary muscle

Lens

Aqueous humour: watery fluid which supports the cornea and the front chamber of the eye.

Cornea: a transparent layer which is responsible for most of the refraction (bending) of light rays from the object to form the inverted, and smaller image on the retina.

Pupil: the circular opening which lets light into the eye. Appears 'black' because the choroid is visible through it.

Iris: the 'coloured' part of the eye which may expand and contract to control the amount of light which enters the eye – this is a **reflex action**.

High light intensity: circular muscles of iris contract and the pupil is reduced in size – **less light may enter and retina is protected from bleaching**.

Low light intensity: radial muscles of iris contract and the pupil is opened wider – **more light may enter and reach retina**.

The brain is an integrator

— it is able to accept sensory information from a number of sources, compare it with previous experience (learning) and make sure that the appropriate actions are initiated.

DON'T FORGET

- in addition to its being a centre of learning the brain is the relay centre for **cranial reflexes** such as blinking.
- **conditioned reflexes** involve setting up nerve pathways in the brain which link a substitute stimulus to a certain response e.g. hearing a lunch bell may set off the production of saliva.
- the evolutionary advantages of mammals may be partly due to a constant body temperature allowing the brain to 'work' for longer periods so that learning is more successful.

DRUGS AND BRAIN FUNCTION

Alcohol has a progressive effect on the brain

- 1–2 units release inhibitions by affecting emotional centres in the forebrain
- 5–6 units affect co-ordinated movements by influencing motor areas of the cerebral cortex
- 7–8 units cause stupor and insensitivity to pain as sensory areas, including the visual centre, are impaired
- more than 10 units may be fatal as the vital centres of the medulla and hypothalamus are severely inhibited.

Heroin can cause euphoria and insensitivity to pain as pain and emotion centres are inhibited.

L.S.D. may cause hallucinations by altering the balance of brain neurotransmitters and causing loss of the medulla's ability to 'filter' information reaching the cerebral cortex.

Cerebral cortex: has motor areas to control voluntary movement, sensory areas which interpret sensations and association areas to link the activity of motor and sensory regions. The centre of intelligence, memory, language and consciousness.

Meninges: membranes which line the skull and cover the brain. They help to protect and nourish the brain tissues, but may be infected either by a virus or a bacterium to cause the potentially fatal condition **meningitis**.

Skull: the cranium is a bony 'box' which encloses and protects the brain.

Visual centre: this area of the cerebral cortex
a interprets impulses along the optic nerve i.e. responsible for vision
b has the connector neurones for both accommodation and the pupil reflex.

Cerebellum: co-ordinates movement using sensory information from position receptors in various parts of the body. Helps to maintain posture using sensory information from the inner ear. Can control learned sequences of activity involved in dancing, athletic pursuits and in the playing of musical instruments.

Forebrain: here many emotions are localised, and damage to this area may cause aggression, apathy, extreme sexual behaviour and other emotional disturbances.

Hypothalamus: contains centres which control thirst, hunger and thermoregulation.

Pituitary gland: is a link between the central nervous system and the endocrine system. Secretes a number of hormones, including follicle stimulating hormone which regulates development of female gametes, and anti-diuretic hormone which controls water retention by the kidney.

Medulla: the link between the spinal cord and the brain, and relays information between these two structure. Has a number of reflex centres which control
a vital reflexes which regulate heartbeat, breathing and blood vessel diameter
b non-vital reflexes which co-ordinate swallowing, salivation, coughing and sneezing.

Spine (vertebral column): composed of 33 separate vertebra which surround and protect the spinal cord; between each pair of vertebrae two **spinal nerves** carry sensory information into the spinal cord and motor information out of it. Dislocation of the vertebrae may compress the spinal nerves, causing great pain, or even crush the spinal cord, leading to paralysis.

Blood sugar regulation is another example of **homeostasis**.

WHAT IS BLOOD SUGAR LEVEL?

Glucose is the cells' main source of energy, and it must always be available to them...

GLUCOSE + OXYGEN → CARBON DIOXIDE + WATER + ENERGY ⇒ Cells may carry out WORK

...so that the body keeps a constant amount of glucose in the blood. This is the **blood sugar level** and is usually maintained at about 1 mg of glucose per cm³ of blood.

WHAT IS DIABETES?

Diabetes is a condition in which there are higher than normal blood glucose concentrations

- usually the result of the failure of the pancreas to secrete enough insulin
- symptoms include
  excessive thirst, hunger or urine production
  sweet smelling breath
  high 'overflow' of glucose into urine (test with **clinistix**)
- long-term effects if untreated include
  premature ageing
  cataract formation
  hardening of arteries
  heart disease
- treatment is by regular injection of pure insulin - much of this is now manufactured by **genetic engineering**.
- mild diabetes can be controlled by adjusting diet to reduce sugar intake.

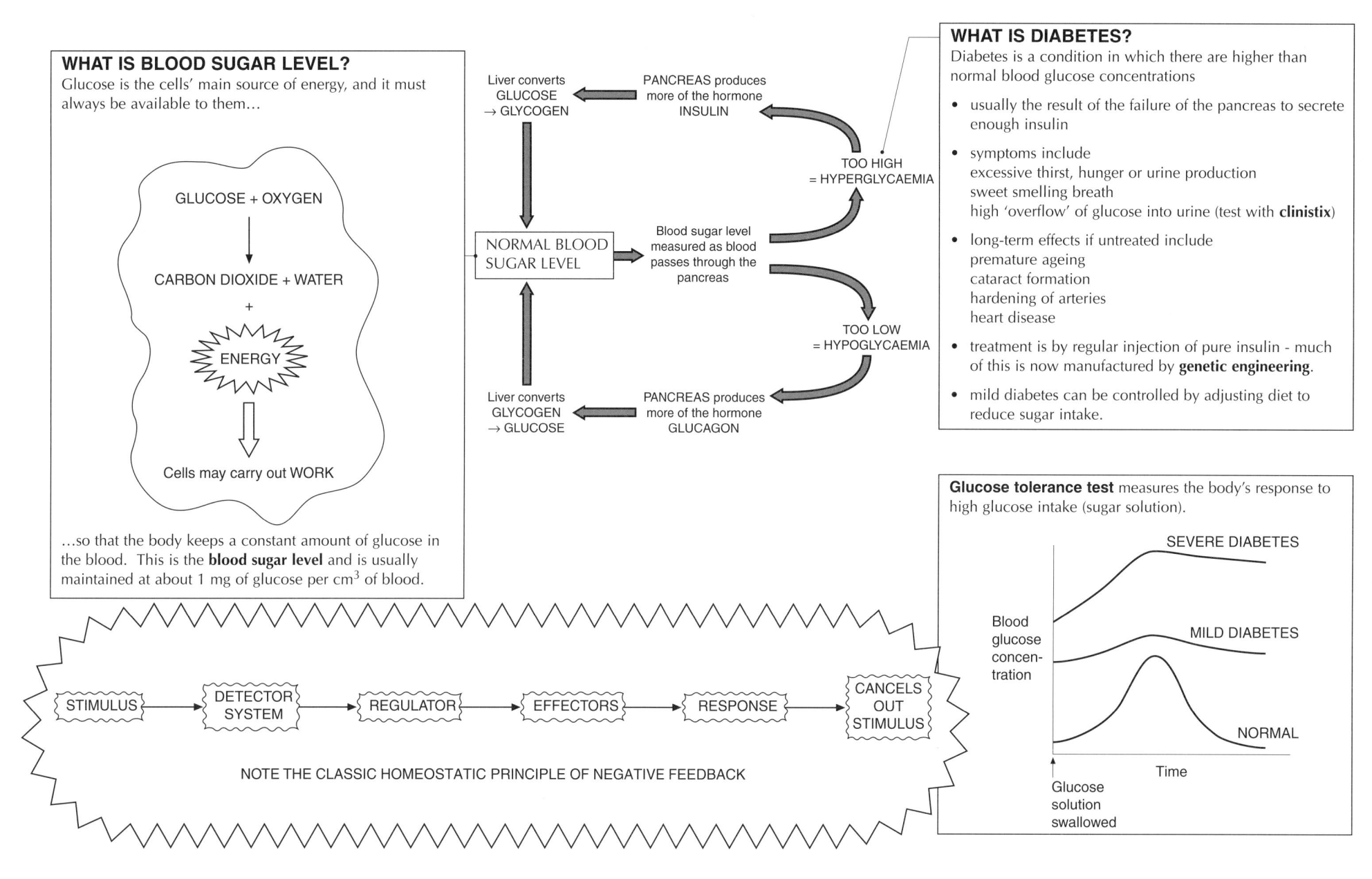

TOO HIGH = HYPERGLYCAEMIA

TOO LOW = HYPOGLYCAEMIA

PANCREAS produces more of the hormone INSULIN
Liver converts GLUCOSE → GLYCOGEN

Blood sugar level measured as blood passes through the pancreas

NORMAL BLOOD SUGAR LEVEL

PANCREAS produces more of the hormone GLUCAGON
Liver converts GLYCOGEN → GLUCOSE

Glucose tolerance test measures the body's response to high glucose intake (sugar solution).

SEVERE DIABETES / MILD DIABETES / NORMAL

Blood glucose concentration vs Time; Glucose solution swallowed

STIMULUS → DETECTOR SYSTEM → REGULATOR → EFFECTORS → RESPONSE → CANCELS OUT STIMULUS

NOTE THE CLASSIC HOMEOSTATIC PRINCIPLE OF NEGATIVE FEEDBACK

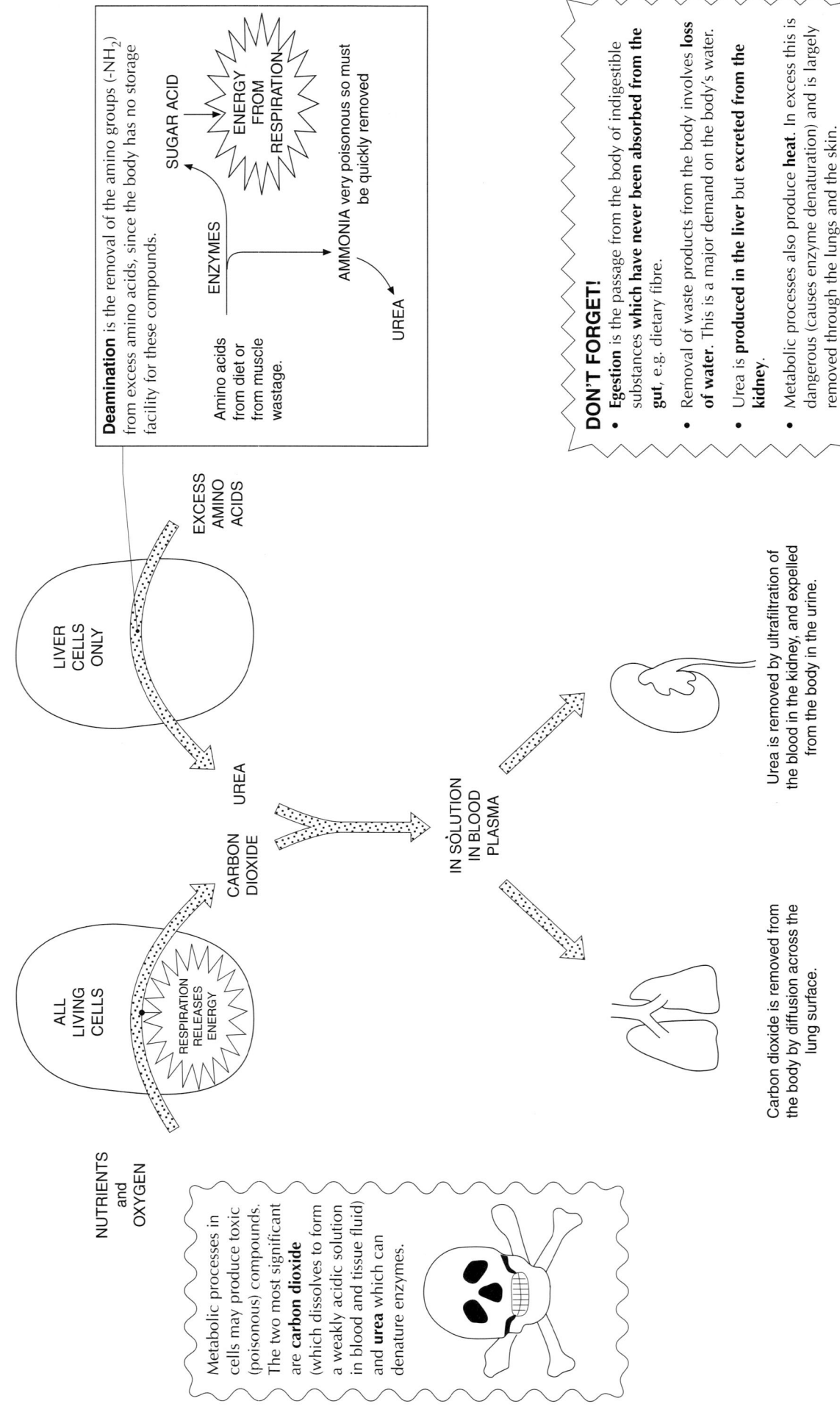

Kidney structure and function

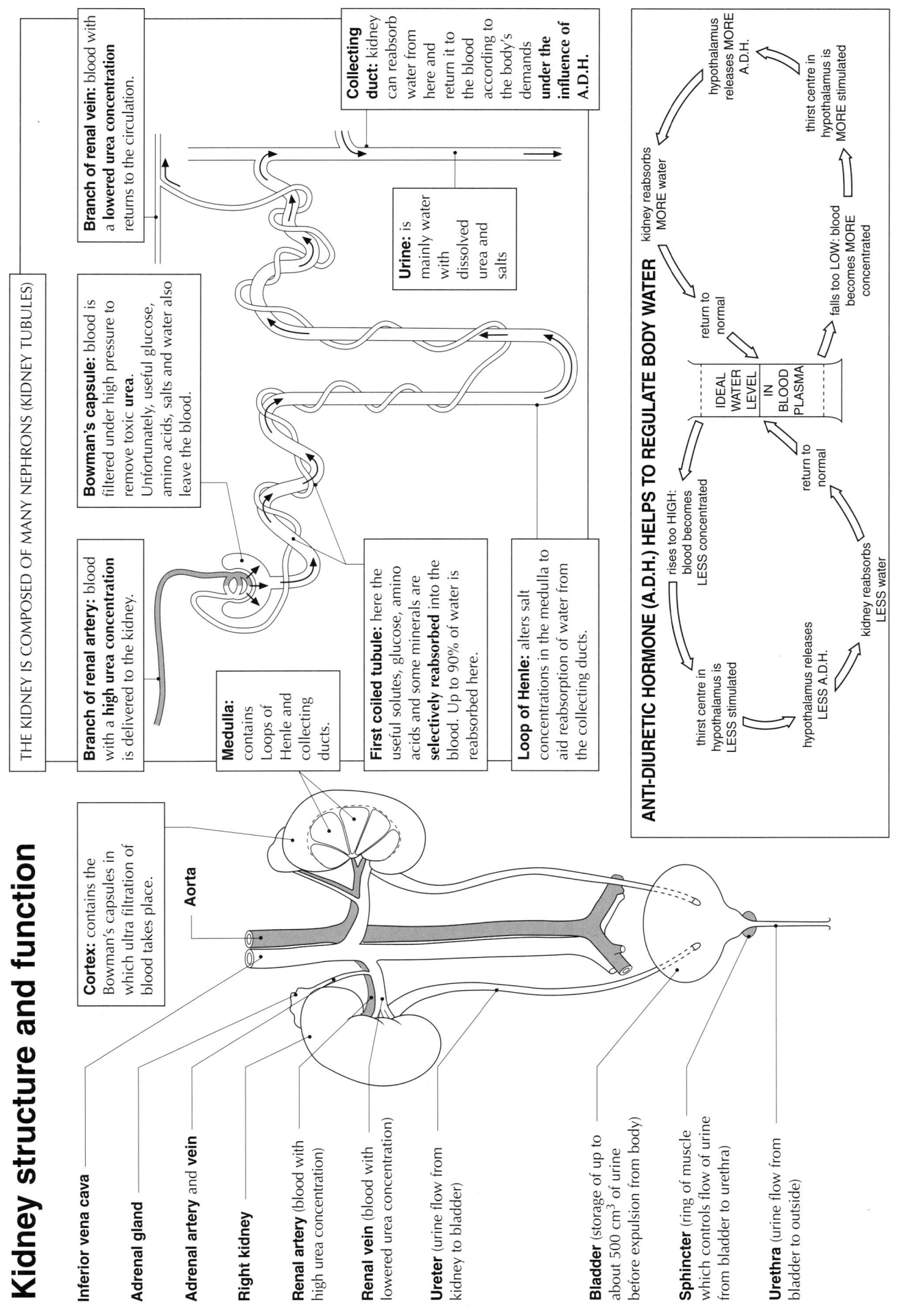

Kidney structure and function 27

Kidney failure:
if one or both kidneys fail then dialysis is used or a transplant performed to keep urea and solute concentration in the blood constant.

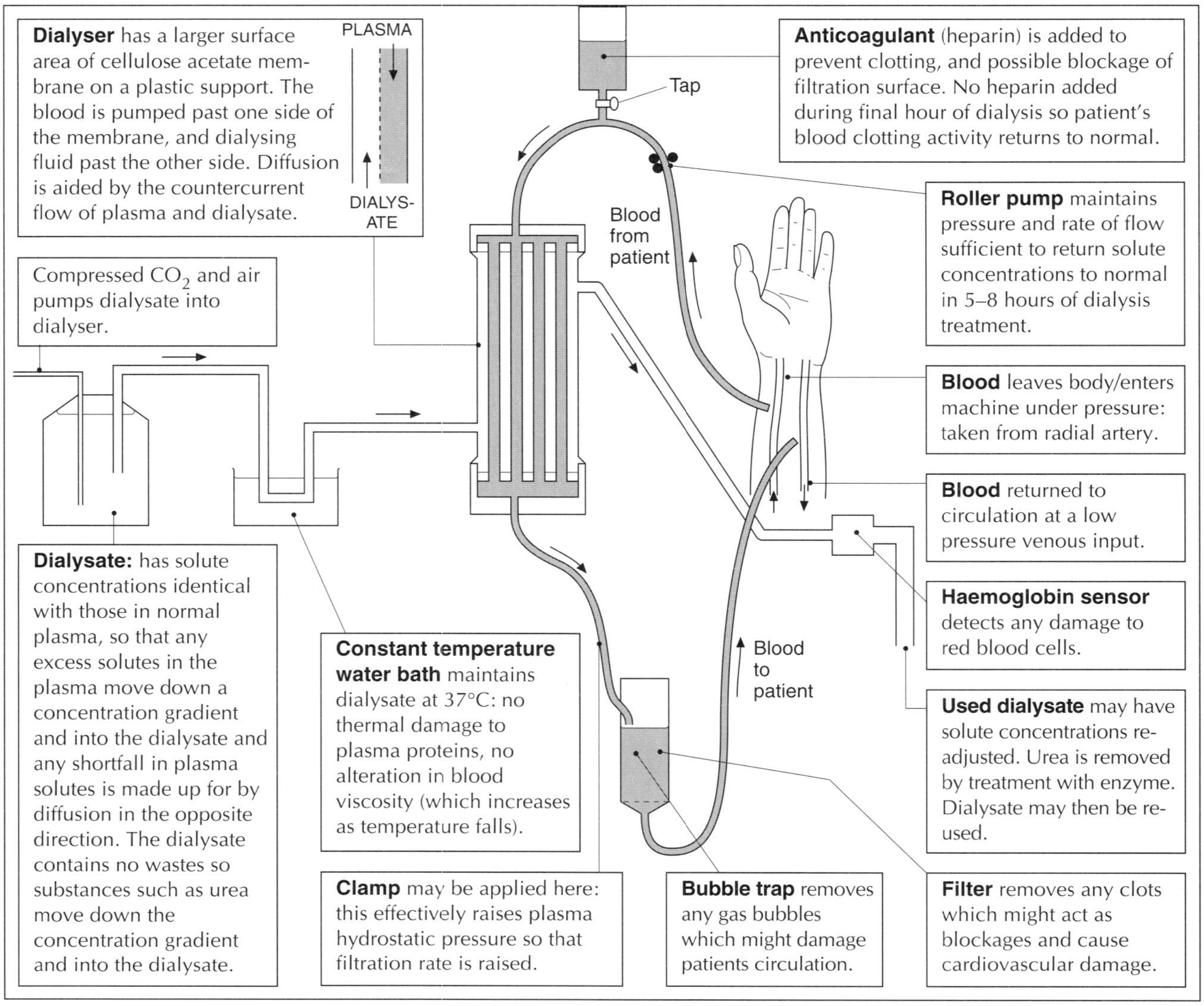

Dialyser has a larger surface area of cellulose acetate membrane on a plastic support. The blood is pumped past one side of the membrane, and dialysing fluid past the other side. Diffusion is aided by the countercurrent flow of plasma and dialysate.

PLASMA ↓
DIALYSATE ↑

Compressed CO_2 and air pumps dialysate into dialyser.

Dialysate: has solute concentrations identical with those in normal plasma, so that any excess solutes in the plasma move down a concentration gradient and into the dialysate and any shortfall in plasma solutes is made up for by diffusion in the opposite direction. The dialysate contains no wastes so substances such as urea move down the concentration gradient and into the dialysate.

Constant temperature water bath maintains dialysate at 37°C: no thermal damage to plasma proteins, no alteration in blood viscosity (which increases as temperature falls).

Clamp may be applied here: this effectively raises plasma hydrostatic pressure so that filtration rate is raised.

Tap

Blood from patient

Anticoagulant (heparin) is added to prevent clotting, and possible blockage of filtration surface. No heparin added during final hour of dialysis so patient's blood clotting activity returns to normal.

Roller pump maintains pressure and rate of flow sufficient to return solute concentrations to normal in 5–8 hours of dialysis treatment.

Blood leaves body/enters machine under pressure: taken from radial artery.

Blood returned to circulation at a low pressure venous input.

Haemoglobin sensor detects any damage to red blood cells.

Blood to patient

Used dialysate may have solute concentrations re-adjusted. Urea is removed by treatment with enzyme. Dialysate may then be re-used.

Bubble trap removes any gas bubbles which might damage patients circulation.

Filter removes any clots which might act as blockages and cause cardiovascular damage.

KIDNEY TRANSPLANTATION

may be necessary as renal dialysis is inconvenient for the patient and costly.

Kidney transplants have a high success rate because:

1. the vascular connections are simple
2. live donors may be used, so very close blood group matching is possible
3. because of 2 there are fewer immuno-suppression-related problems in which the body's immune system reacts against the new kidney.

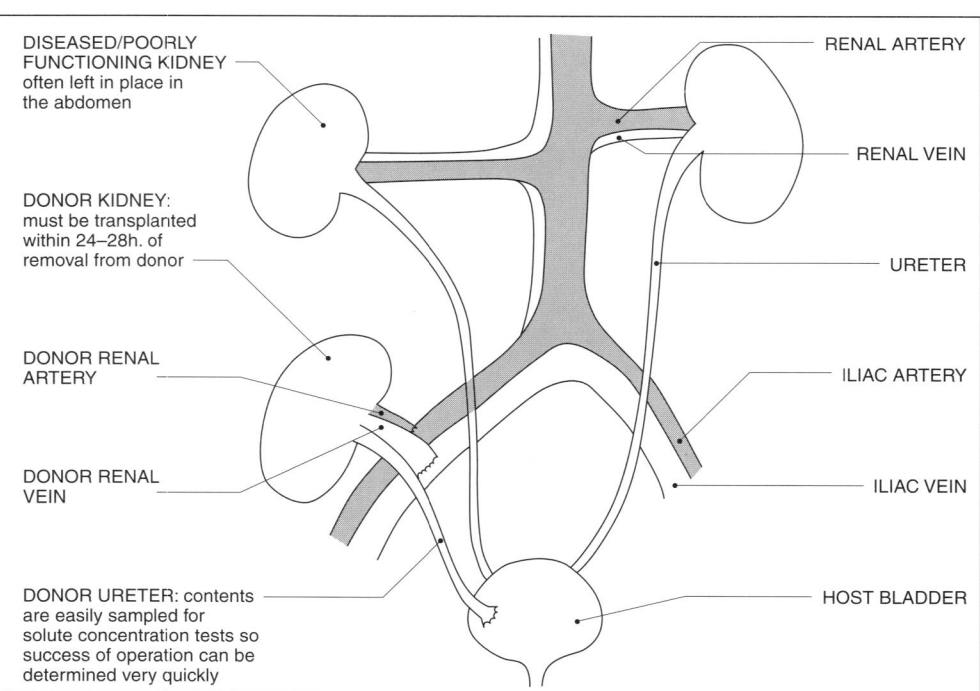

DISEASED/POORLY FUNCTIONING KIDNEY often left in place in the abdomen

DONOR KIDNEY: must be transplanted within 24–28h. of removal from donor

DONOR RENAL ARTERY

DONOR RENAL VEIN

DONOR URETER: contents are easily sampled for solute concentration tests so success of operation can be determined very quickly

RENAL ARTERY
RENAL VEIN
URETER
ILIAC ARTERY
ILIAC VEIN
HOST BLADDER

Temperature control in endothermic organisms

is an example of **negative feedback**. Any deviation from optimum conditions sets mechanisms in motion to cancel out the change.

The **temperature control centre** is located in the **hypothalamus** in the floor of the brain. It receives incoming messages about the body temperature and makes certain that the appropriate corrective mechanisms operate to cancel out any deviations.

Skin increases heat loss

Sweat is secreted and **evaporation** consumes heat from body.

Vasodilation of surface capillaries allows blood to **radiate** heat away from body.

Relaxation of hair erector muscles allows hair to lie flat against skin to permit maximum heat loss by convection.

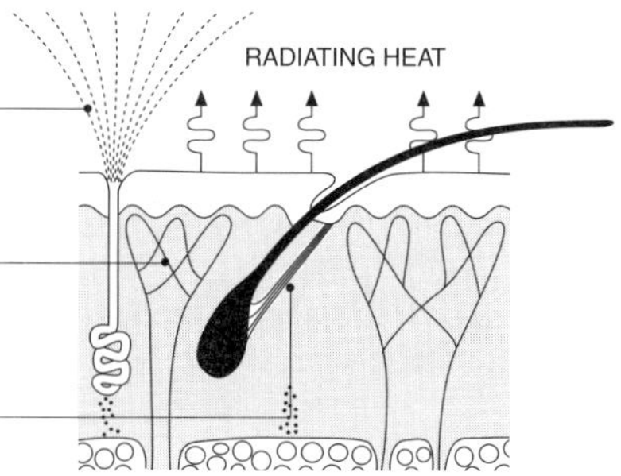

RADIATING HEAT

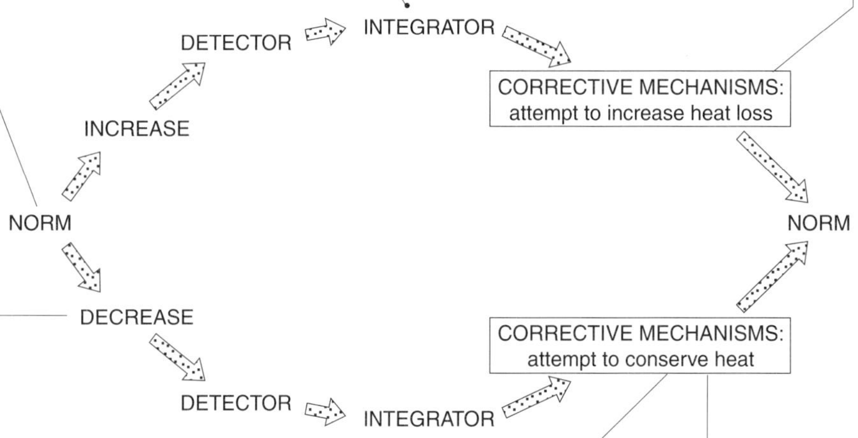

During **fever**, the body's temperature may be reset 2–3°C higher: the higher temperatures denature proteins in the infecting bacteria more rapidly than they affect human proteins.

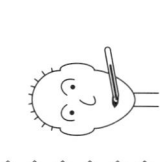

CORRECTIVE MECHANISMS: attempt to increase heat loss

↑ INTEGRATOR

↑ DETECTOR

↑ INCREASE

NORM

↓ DECREASE

↓ DETECTOR

↓ INTEGRATOR

CORRECTIVE MECHANISMS: attempt to conserve heat

Skin reduces heat loss

Contraction of hair erector muscles traps a layer of still air which reduces heat losses by **convection**.

Vasoconstriction of surface capillaries shunts blood away from skin and so reduces heat losses by **radiation**.

Sweat glands do not secrete sweat so there is no consumption of body heat in **evaporation**.

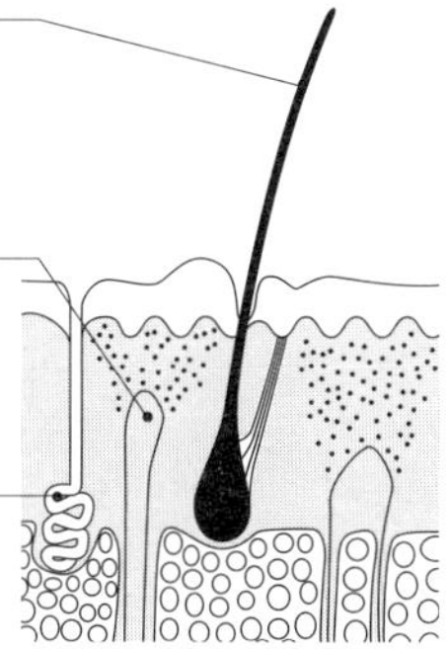

The **optimum temperature** for the body's activities is about 37°C. This ensures that

- enzymes work rapidly but are not denatured
- cell membranes keep their structure
- blood does not become too viscous to pass through narrow capillaries.

Deviations from the norm are detected
a. by **hot** and **cold sensors** in the skin
b. by sensors in the hypothalamus which measure blood temperature.

Humans may also

shiver – muscular activity generates heat

add extra clothing – insulation reduces heat loss

eat more – eating stimulates heat production by respiration…

…or **turn on the central heating**!

(humans can control their **external** environment too!)

Hypothermia ("low temperature")

- occurs if body's core temperature falls below 35°C
- causes irrational behaviour, brain damage, circulatory problems
- elderly people are very susceptible because they may be immobile
- children are very susceptible because they have a high surface area to volume ratio.

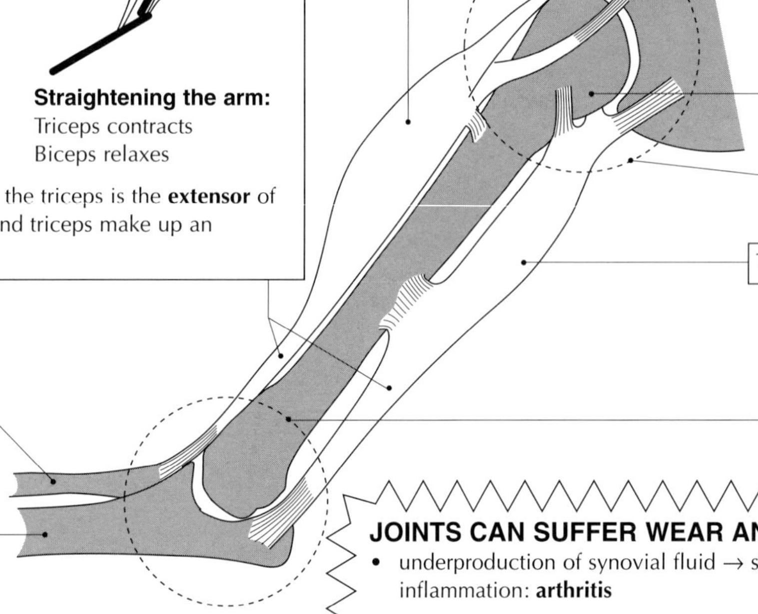

Muscle-bone machines
are responsible for movement.

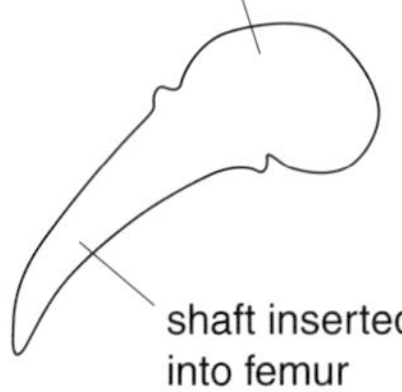

CONTRACTION OF MUSCLE REQUIRES
- the **supply** of **glucose** and **oxygen** to release energy, in the form of ATP, by respiration
- the **removal** of **carbon dioxide** and **heat** (if respiration becomes **anaerobic** there must also be the removal of **lactic acid** or cramp may result)

These requirements are satisfied by the development of a system of capillary beds in the muscle together with their arteries and veins.

- a **stimulus** in the form of a **nervous impulse** delivered from the end plates of **motor neurones**.

Tendons connect muscle to bone. They are **inelastic** so that muscle contraction can be converted into movement of bone.

Antagonistic pairs of muscles are necessary for controlled movement at a joint. Muscles may only exert a force by **contraction**. To reverse a muscular movement therefore requires contraction of an opposing (antagonistic) muscle.

Human forelimb

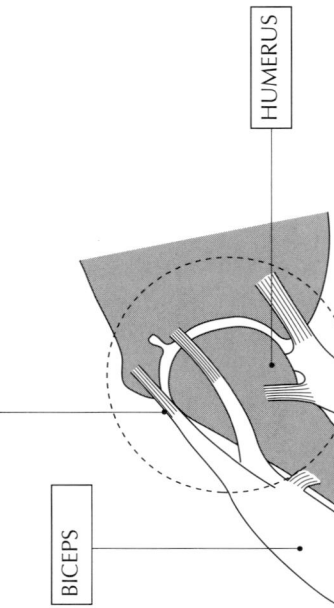

Bending the arm:
Biceps contracts
Triceps relaxes

Straightening the arm:
Triceps contracts
Biceps relaxes

So, the biceps is a **flexor** and the triceps is the **extensor** of the elbow joint, and biceps and triceps make up an **antagonistic pair**.

HUMERUS

BICEPS

TRICEPS

RADIUS

ULNA

Synovial joints are freely-moving because of the arrangement of the synovial membranes/cartilage between the ends of adjacent bones.

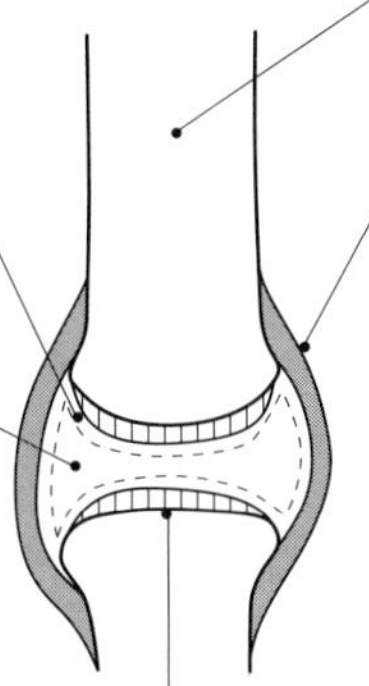

Synovial membrane secretes the synovial fluid

Synovial fluid looks and feels like egg-white. It lubricates the joint and contains phagocytic cells which remove debris resulting from wear and tear in the joint.

Bone

Ligaments are elastic but strong. They **allow movement** at the joint but help to **prevent dislocation**. Ligaments connect bone to bone.

Cartilage covers the ends of the two bones which meet at a joint. It **reduces friction** between the bones and helps to **absorb shock** if the joint is compressed.

Ball-and-socket joints (e.g. hip and shoulder) allow free movement in many planes.

Hinge joints (e.g. elbow, knuckles) allow free movement in one plane only.

JOINTS CAN SUFFER WEAR AND DAMAGE
- underproduction of synovial fluid → stiffness and inflammation: **arthritis**
- dislocation may occur if ligaments are torn of overstretched
- cartilage may be displaced following impact on a joint. The cartilage fragments may impede movement and may need to be removed.
- **replacement joints** may be used if original is damaged/severely arthritic. These are made from stainless steel (strong, non-toxic, non-corrosive) and may be coated with 'teflon' on the working surface.

ball to fit hip joint

shaft inserted into femur

OTHER FUNCTIONS OF THE BONY SKELETON
- **support** - air itself is not dense enough to support the body of a land mammal
- **protection** - soft tissues may be enclosed inside a bony case e.g. brain inside the skull.
- **synthesis** - red and white blood cells are produced in bone marrow inside long bones such as the femur.

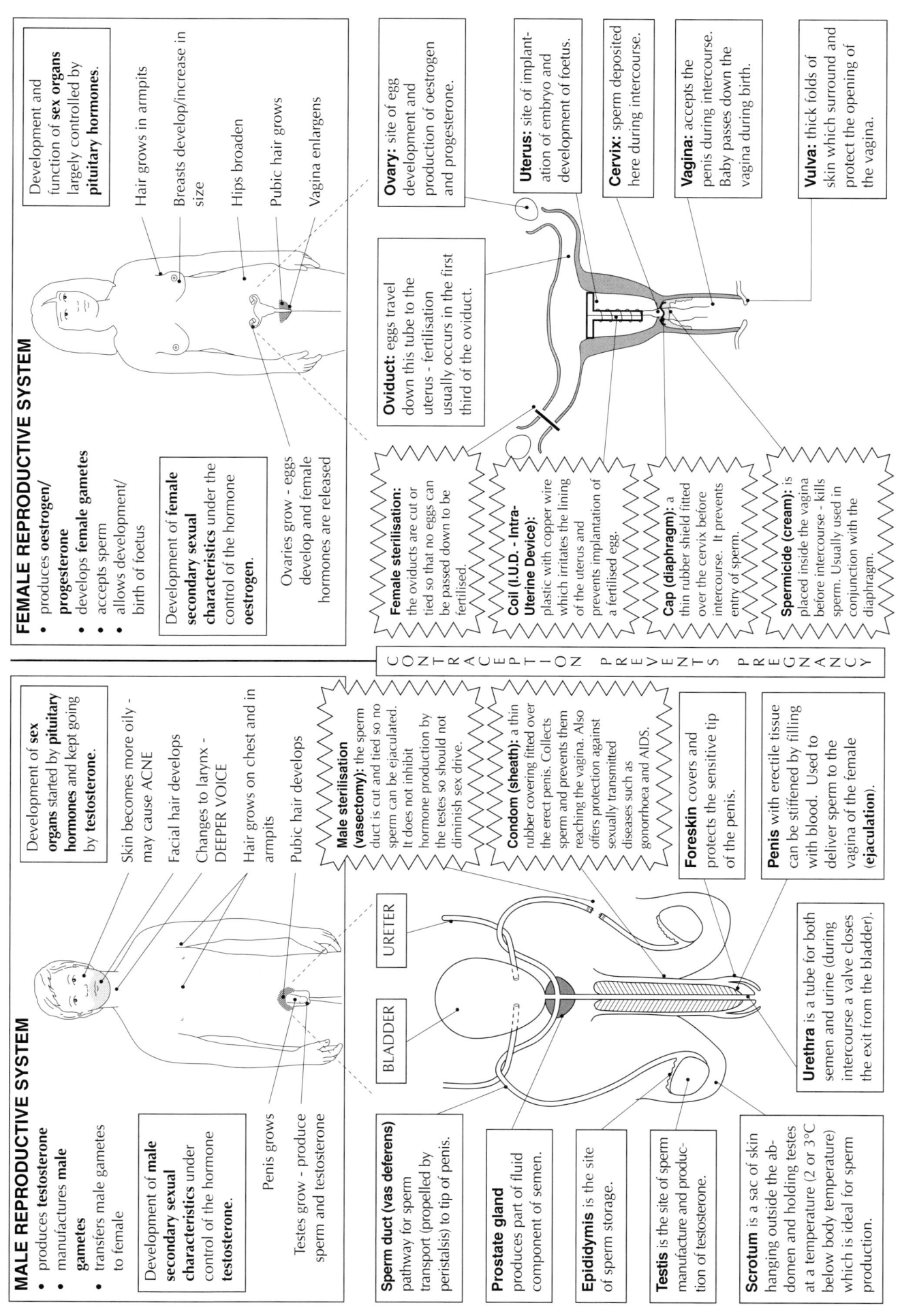

The menstrual cycle

The Menstrual cycle is a regular series of changes to the female reproductive system in preparation for fertilisation and pregnancy. It is controlled by **hormones** from the **pituitary gland** and the **ovary**.

In humans the cycles of the two ovaries are out of phase, so **each ovary** ovulates every 56 days but **each female** ovulates (releases an egg) every 28 days.

Separation of uterus lining - blood and fragments of tissue leave the body via the vagina. This monthly 'period' occurs only in primates such as humans.

A TYPICAL MENSTRUAL CYCLE
- Stage 1 MENSTRUATION
- Stage 2 REPAIR PHASE (of uterus lining)
- Stage 3 RECEPTIVE PHASE
- Stage 4 PRE-MENSTRUAL PHASE

The uterus lining begins to degenerate **unless embryo implantation** has occurred when **progesterone** (from the corpus luteum) keeps the lining intact to begin pregnancy.

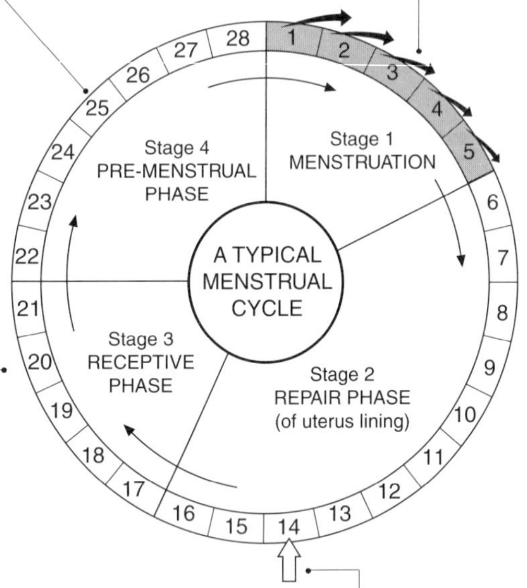

Uterus lining and its blood vessels are well-developed to receive an embryo. This optimum set of conditions remains for 6-7 days after ovulation.

Ovulation, the release of the egg from the Graafian follicle into the oviduct. This is stimulated by **luteinising hormone** and the release of fluid and an associated 'blip' in body temperature (see right) means that some human females are aware that ovulation has occurred.

Contraception and the menstrual cycle

- **The pill:** an oral dose of one or both of the hormones **oestrogen** and **progesterone** which acts as a **feedback inhibitor** of the release of luteinising hormone by the pituitary gland. Thus **ovulation cannot occur** and **no pregnancy can result**.
- **The rhythm method** assumes that the time up to two days before ovulation and from four days after ovulation are **safe**, i.e. there should be no eggs available to be fertilised. This method is unreliable.

THE MENSTRUAL CYCLE IS CONTROLLED BY HORMONES

Brain - receives information from other parts of the body, processes it and then 'instructs' pituitary gland.

Pituitary gland - releases hormones which control activity of the ovary.

Luteinising hormone - stimulates release of mature ovum from ovary.

Luteinising hormone - stimulates development of corpus luteum from the remains of the follicle.

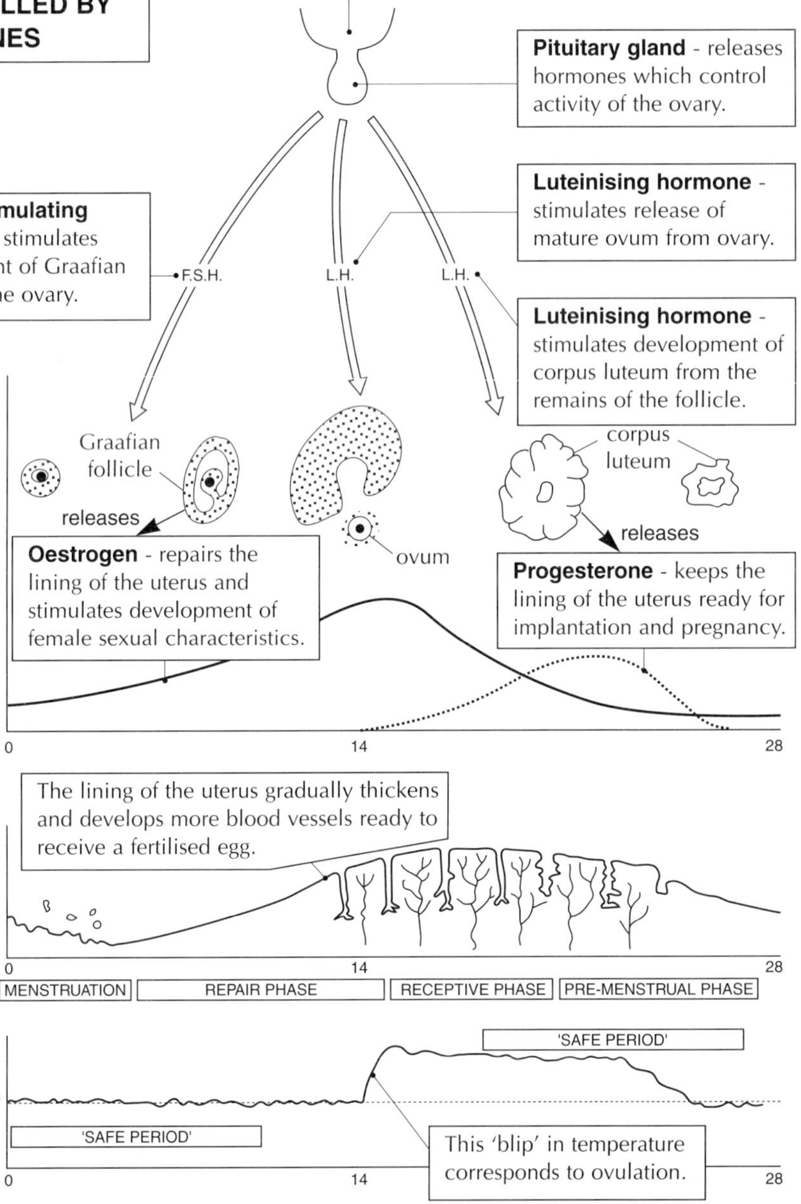

Progesterone - keeps the lining of the uterus ready for implantation and pregnancy.

Follicle stimulating hormone - stimulates development of Graafian follicle in the ovary.

Oestrogen - repairs the lining of the uterus and stimulates development of female sexual characteristics.

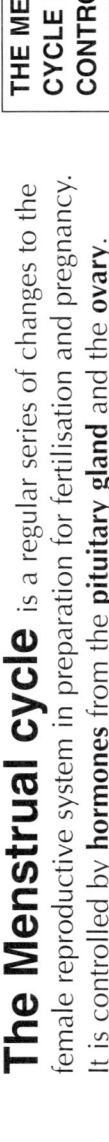

The lining of the uterus gradually thickens and develops more blood vessels ready to receive a fertilised egg.

MENSTRUATION | REPAIR PHASE | RECEPTIVE PHASE | PRE-MENSTRUAL PHASE

'SAFE PERIOD' — 'SAFE PERIOD'

This 'blip' in temperature corresponds to ovulation.

The placenta

The placenta is responsible for protection and nourishment of the developing foetus.

PLACENTA: this disc-shaped organ has a number of functions:

- exchange of soluble materials between mother and foetus
- physical attachment of the foetus to the wall of the uterus
- protection (1) of foetus from mother's immune system
  (2) against dangerous fluctuations in mother's blood pressure
- hormone secretion. These hormones keep the wall of the uterus in 'pregnancy' state as the corpus luteum breaks down by the 3rd month.

The placenta is lost as the 'afterbirth' following birth of the foetus.

UMBILICAL CORD: contains blood vessels which carry materials which will be / have been exchanged between mother and foetus. The cord connects the foetus to the placenta.

PASSING FROM MOTHER TO FOETUS
- Soluble nutrients e.g. glucose and amino acids
- Oxygen
- antibodies

BUT also
- Viruses e.g. HIV carbon monoxide (e.g. from smoking) nicotine/heroin

AND FROM FOETUS TO MOTHER
- carbon dioxide
- urea

Placental villi - increase the surface area for exchange.

Umbilical artery from foetus to placenta - deoxygenated blood containing waste.

Umbilical vein from placenta to foetus - oxygenated blood cleared of waste.

Counter-current flow system blood in maternal and foetal blood vessels **flows in opposite directions**. This gives the maximum possible area over which concentration gradients favour diffusion.

'Pit' in wall of uterus contains mother's blood. N.B. there is **no direct contact** between maternal and foetal blood.

Wall of uterus: this is very muscular. At full term of the pregnancy a hormone (OXYTOCIN) is secreted from the pituitary gland which makes this muscle contract in a series of waves to expel the foetus. The same hormone is used in the intravenous drip which induces birth when the pregnancy has gone on for too long.

Foetus: this develops from a **zygote**.

ZYGOTE at fertilisation

CELL DIVISION

BALL OF CELLS at implantation

CELL DIVISION AND MOVEMENT

EMBRYO

CELL DIVISION, MOVEMENT AND SPECIALISATION

Recognisably human form by end of eighth week: FOETUS

Amnion: the membrane which encloses the amniotic fluid. This is 'ruptured' just before birth.

Amniotic fluid: protects the foetus against
- mechanical shock
- drying out
- temperature fluctuations

Some of the foetal cells fall off into this fluid and can be collected by **amniocentesis**. The cells can be analysed to detect disease, genetic abnormalities and even the sex of the foetus.

Mucus plug in cervix: protects foetus against possible infection. The plug is expelled just before birth.

GROWTH AND DEVELOPMENT OF THE FOETUS

MASS/g: 1000, 2000, 3000
AGE OF FOETUS/weeks: 0 4 8 12 16 20 24 28 32 36 40

- Internal organs all present
- Foetus now has a good chance of survival if born
- Foetus fully formed, even fingerprints!
- Birth usually occurs at about this time

- Complete period from fertilisation to birth = **gestation period**.
- Cell division converts single cell (zygote) to 30 million in a newborn baby.
- Most rapid growth from the twelfth week. As much as 1500x gain in mass in 20 weeks.

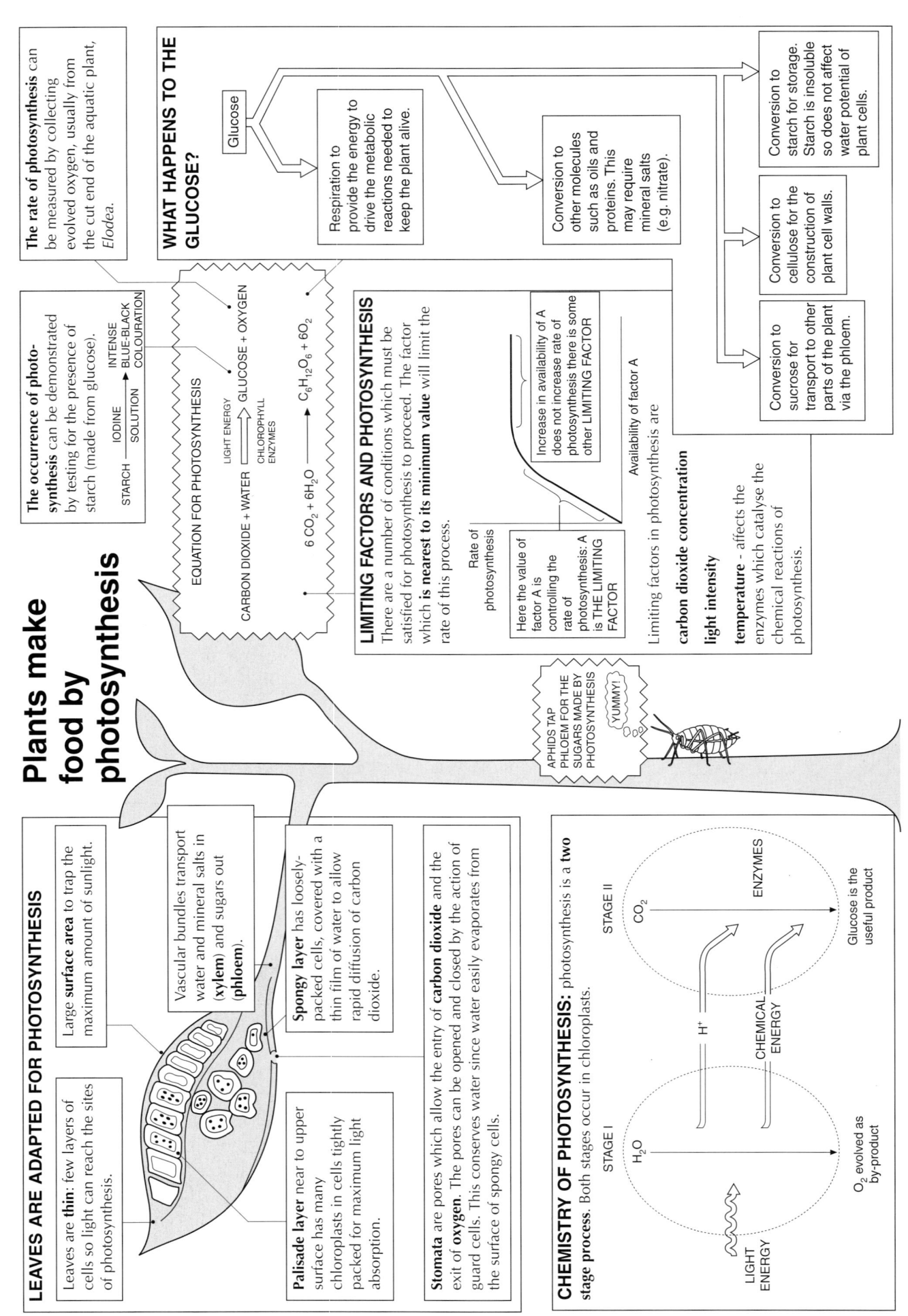

Transport systems in plants

move vital substances from **sources** (sites of uptake or manufacture) to **sinks** (sites of production or storage).

INVESTIGATION OF FUNCTIONS OF PLANT TISSUES

- **Phloem:** aphids (greenfly) can sample contents and radioactive sugars identify tissue as site of sugar transport.
- **Xylem:** water-soluble dyes trace pathway of water movement
- **Roots:** inhibitors of respiration stop active uptake of ions

FRUITS AND GROWING POINTS: these are **sinks** for many nutrients.

Fruits
- demand water for swelling of ovary wall if succulent
- demand sucrose to be converted to starch as an energy store

Growing points
- demand water for cell expansion
- demand sucrose as energy source for cell division
- demand all nutrients as raw materials for cell production.

LEAVES: are both **sinks** and **sources**.

Sinks
- water as a reactant in photosynthesis
- magnesium as a component of the chlorophyll molecule
- sucrose, when leaves are young, as a source of energy and subunits of cellulose.

Sources
- glucose formed during photosynthesis

  CARBON DIOXIDE + WATER
  $\downarrow$
  GLUCOSE + OXYGEN

- useful ions, e.g. magnesium, just before leaf fall.

DIRECTION OF TRANSPORT VARIES WITH THE SEASONS!

Sucrose will be transported **from** stores in the root **to** leaves in Spring, but **to** stores in the root **from** photosynthesising leaves in the Summer and early Autumn.

ROOTS: are both **sinks** and **sources**.

Sinks
- sucrose to supply energy for growth and active uptake of ions from the soil
- sucrose to be converted to starch for storage

Sources
- water, absorbed from the soil solution
- ions, absorbed from soil by active transport
- sucrose, before leaves are capable of photosynthesis, in Spring.

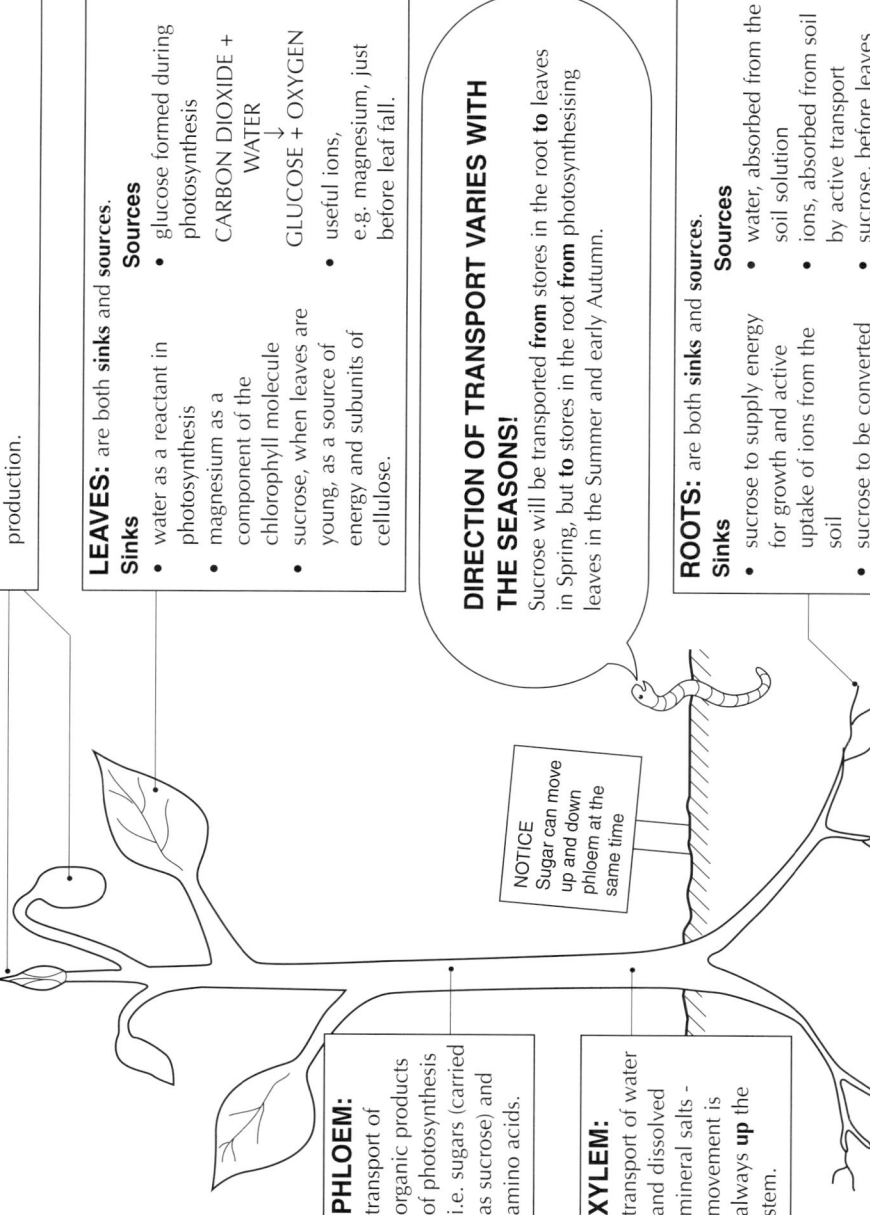

NOTICE Sugar can move up and down phloem at the same time

PHLOEM: transport of organic products of photosynthesis i.e. sugars (carried as sucrose) and amino acids.

XYLEM: transport of water and dissolved mineral salts - movement is always **up** the stem.

STEM:

the position of the vascular bundles (in a ring with soft cortex in the centre) helps to support the stem against sideways forces, e.g. wind.

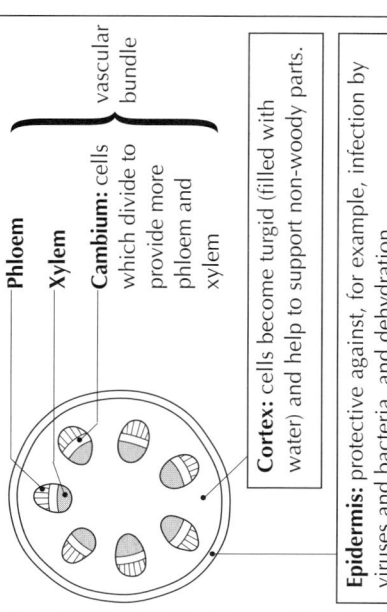

- Phloem
- Xylem
- **Cambium:** cells which divide to provide more phloem and xylem
- vascular bundle

Cortex: cells become turgid (filled with water) and help to support non-woody parts.

Epidermis: protective against, for example, infection by viruses and bacteria, and dehydration.

ROOT:

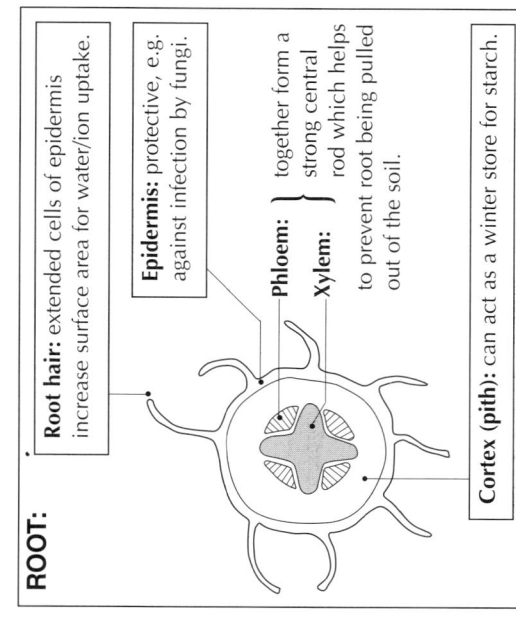

Root hair: extended cells of epidermis increase surface area for water/ion uptake.

Epidermis: protective, e.g. against infection by fungi.

Phloem:
Xylem: together form a strong central rod which helps to prevent root being pulled out of the soil.

Cortex (pith): can act as a winter store for starch.

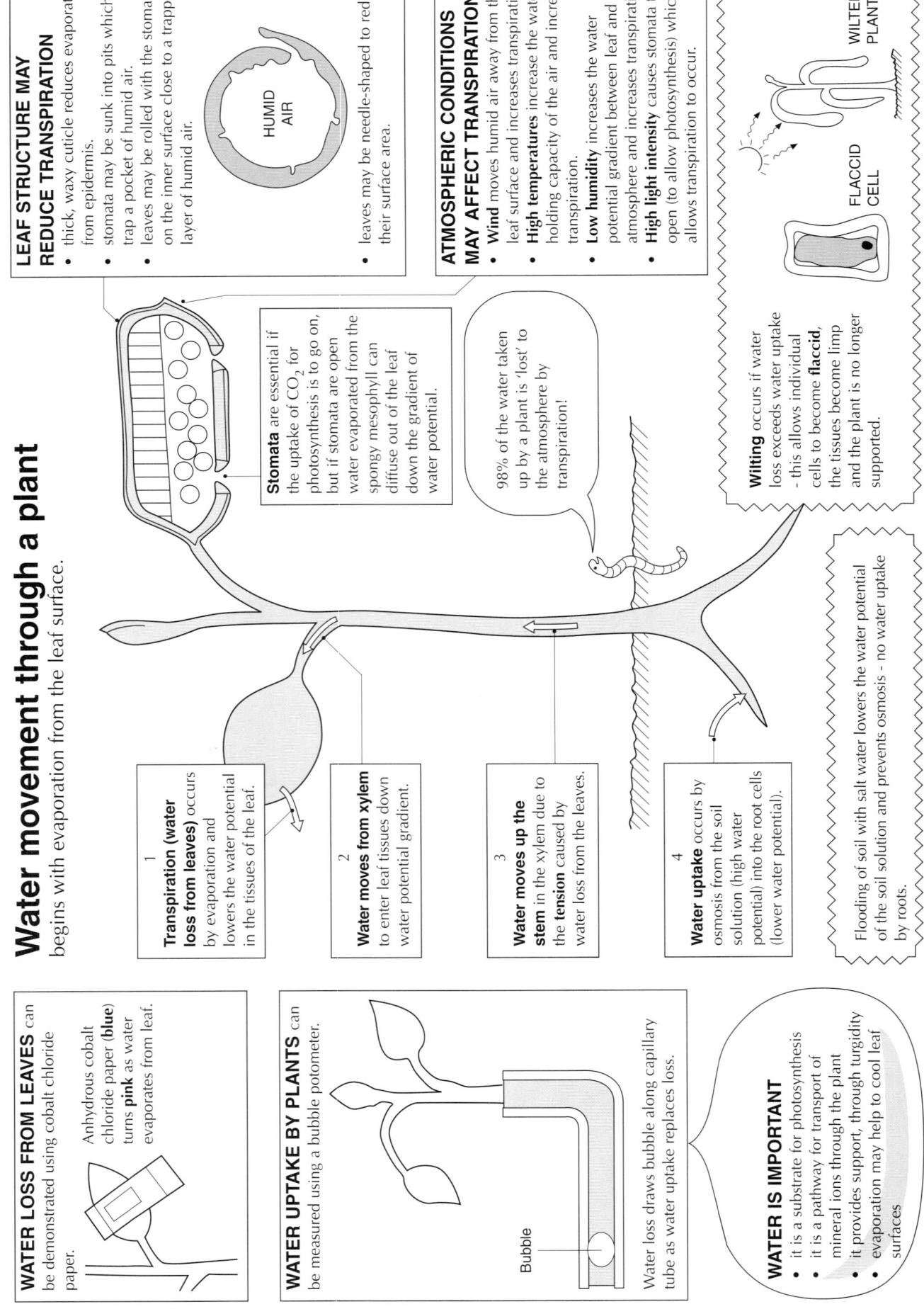

Hormones and minerals affect plant growth and development

MINERALS

MAGNESIUM is absorbed from the soil solution as the ion, Mg^{2+}. It is required for **manufacture of chlorophyll** - absence causes leaves to turn pale yellow and eventually halts photosynthesis.

NITROGEN is absorbed from soil solution as nitrate (NO_3^-) or ammonium (NH_4^+) ions. It is required for **manufacture of proteins, nucleic acids** and **plant hormones**. Absence causes poor growth of all organs, especially leaves. Availability of nitrogen is probably the **major limiting factor** in plant growth in cultivated soils so many farmers add **nitrogen-containing fertilisers** to their land.

WATER CULTURE EXPERIMENTS

allow study of plant mineral requirements in controlled conditions which eliminate the many variables associated with soil as a growth medium.

Cereal seedling is rapid-growing so that effects of mineral deficiency are observed in a short time.

Light-proof cover (1) prevents entry of airborne contaminants and (2) supports seedling in growing position.

Black card or foil cover prevents entry of light so that no aquatic organisms can compete with seedling roots for mineral ions. Cover can be easily removed to examine the growth of the seedling roots.

Glass container can be thoroughly cleaned (using acid) so that no minute traces of minerals remain where they might lead to erroneous results.

Growth/nutrient solution contains mineral ions in previously determined optimum concentrations. Complete (control) solution contains all ions. Test solutions have a single ion omitted. NB to eliminate one ion, a salt is omitted but its 'second' ion is replaced.
e.g. for nitrogen-free solution omit calcium nitrate but increase concentration of another calcium salt. The complete solution is often called Knop's solution.

Aeration has two functions:
1. mixing of solution so that no stagnation occurs
2. oxygenation so that aerobic respiration may provide energy for active uptake of ions from solution.

HORMONES

SHOOTS ARE POSITIVELY PHOTOTROPIC -
they show growth movements towards light.

- **Mechanism:** the plant growth substance called **auxin** is inhibited by light. It therefore accumulates on the shaded side of shoots **causing faster growth on the shaded side** and thus **'bending' towards light.**

  Auxin produced at shoot tip diffuses and accumulates on shaded side

  Auxin is inactivated on 'light' side of shoot

  LIGHT

- **Benefit:** leafy shoots tilt towards light to improve rate of photosynthesis and therefore increase rate of growth.

PLANT HORMONES HAVE COMMERCIAL USES

Plant hormones can be sprayed onto flowers to mimic fertilisation and increase the chance of co-ordinated development of fruits.

Plant hormones can act as selective weedkillers since they disrupt the growth pattern of weed species at a concentration which does not affect cereal crops.

At the correct concentration plant hormones can stimulate root formation in cuttings.

ROOTS ARE POSITIVELY GEOTROPIC -
they show growth movements towards gravity.

- **Mechanism:** gravity causes **auxin** to accumulate on lower side of root. Unlike shoot growth, root growth is **inhibited** by this concentration of auxin so that **lower side grows more slowly and root 'bends' towards gravity.**

  Auxin produced at root tip is affected by gravity and accumulates on lower side of root.

- **Benefit:** root grows into soil to anchor plant firmly and improve uptake of minerals and water.

Flower structure is adapted to pollination

POLLINATION is the transfer of pollen from the stamen to the stigma.

Cross-pollination involves pollen transfer between different flowers of the same species. It involves genetic variation but may be risky because it requires a **vector** (a carrier).

Self-pollination involves pollen transfer from stamen to stigma of the same flower. It is less risky but it limits the chances of genetic variation.

Wind-pollinated species usually occur in dense groups e.g. the grasses.

PETALS	Dull in colour. Small, or even absent to reduce obstruction of pollen access to stigma.
STAMENS	Long and flexible so that pollen may be easily released.
STIGMA	Long and feathery, giving a large surface area to receive pollen.
POLLEN	Small, dry, enormous quantities.

Insect-pollinated species are usually solitary or in small groups.

PETALS	Large, brightly-coloured, may be scented and/or have guidelines. Base may produce attractive **nectar**.
STAMENS	Short and stiff to brush pollen against body of visiting insect.
STIGMA	Held inside petals to ensure contact with body of visiting insect.
POLLEN	Large, sticky, small amounts.

Bees are adapted to feed on pollen and nectar. They pollinate flowers when feeding.

- Antennae - detect scent of flower.
- Mouthparts can suck nectar from flower.
- Eyes - detect colour of flower.
- Wings permit movement between flowers.
- Hairy abdomen may hold 'sticky' pollen.
- Pollen brushes and baskets on third pair of legs collect pollen for larvae in hive.

CARPEL: the female part of the flower.

Stigma: surface on which pollen grains, containing male gametes, may be deposited.

Style: stalk which holds stigma in prominent position and down which pollen tube may grow.

Ovary: contains the ovule which encloses the female gamete. Ovary wall may become part of the fruit.

STAMEN: the male part of the flower.

Anther: produces pollen grains, containing male gametes, within the pollen sacs.

Filament: stalk which holds anther in position to release pollen.

PETALS: play an important part in pollination (see opposite).

SEPALS: protect the other floral parts against drying out and fungal attack. There are the same number as there are petals, and they are usually green in colour.

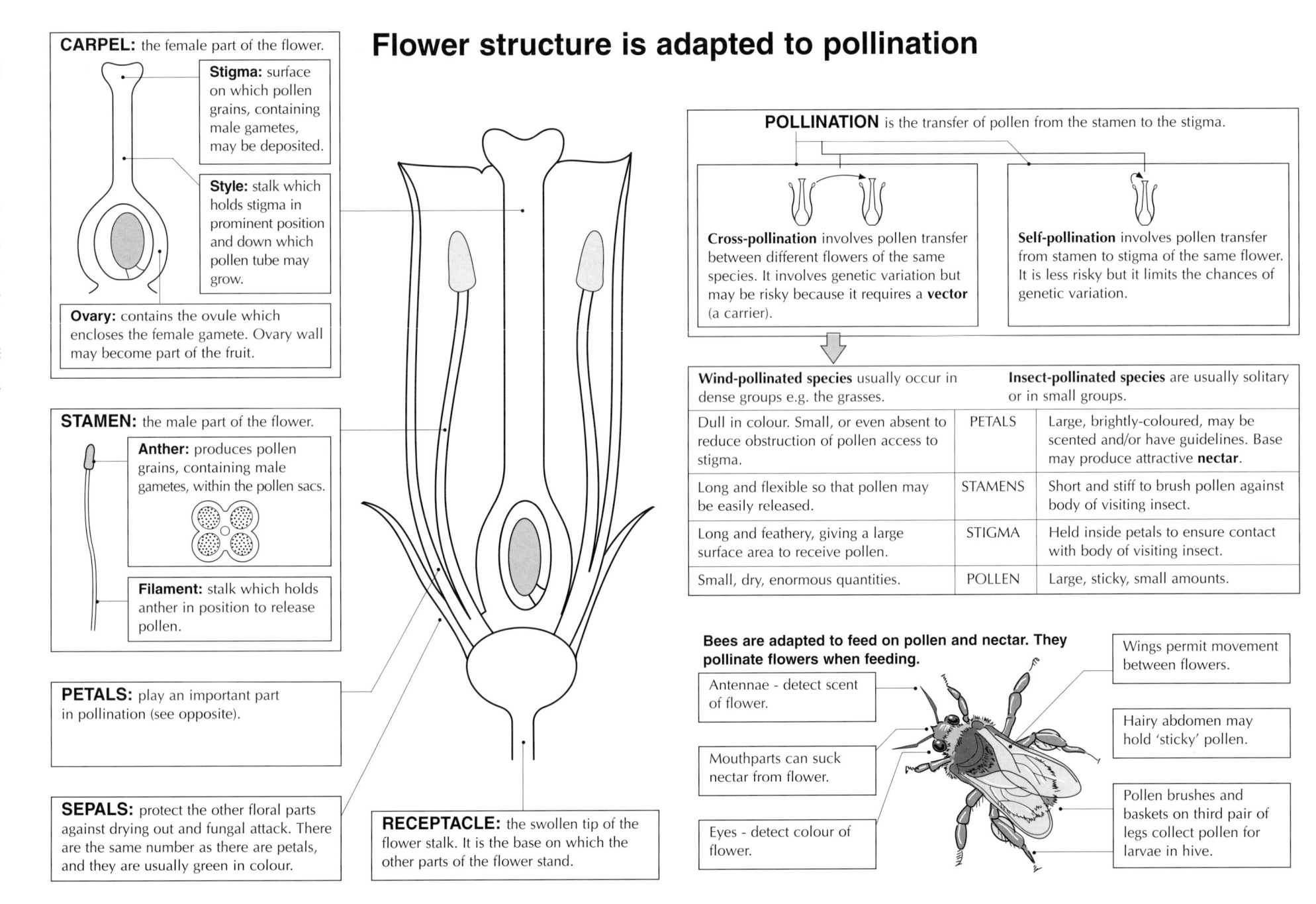

RECEPTACLE: the swollen tip of the flower stalk. It is the base on which the other parts of the flower stand.

Fertilisation, fruits and seed dispersal

Testa (seed coat): prevents drying out of the embryo.

Cotyledons: form the food store for the embryo - may be **one** in monocotyledons like grasses, or **two** in the dicotyledons like peas.

Micropyle: allows entry of water as seed begins to germinate.

The CARPEL → FRUIT with the OVULE → SEED

Embryo: following germination this grows and develops into a new young plant.

The fruit may be **adapted to aid dispersal of seeds**

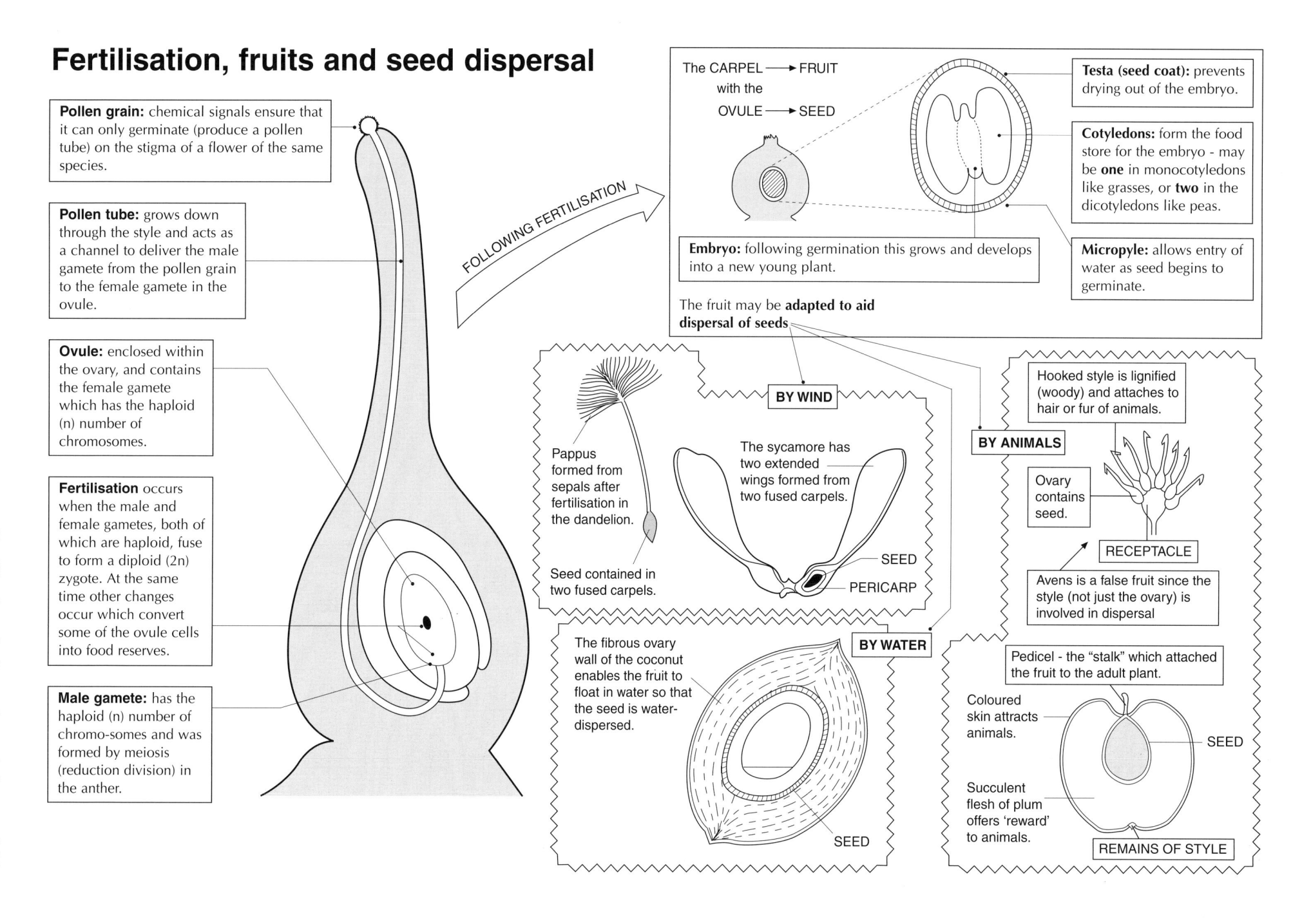

BY ANIMALS

Hooked style is lignified (woody) and attaches to hair or fur of animals.

Ovary contains seed.

RECEPTACLE

Avens is a false fruit since the style (not just the ovary) is involved in dispersal

SEED
REMAINS OF STYLE

Pedicel - the "stalk" which attached the fruit to the adult plant.

Coloured skin attracts animals.

Succulent flesh of plum offers 'reward' to animals.

BY WIND

The sycamore has two extended wings formed from two fused carpels.

SEED
PERICARP

Pappus formed from sepals after fertilisation in the dandelion.

Seed contained in two fused carpels.

BY WATER

SEED

The fibrous ovary wall of the coconut enables the fruit to float in water so that the seed is water-dispersed.

FOLLOWING FERTILISATION

Pollen grain: chemical signals ensure that it can only germinate (produce a pollen tube) on the stigma of a flower of the same species.

Pollen tube: grows down through the style and acts as a channel to deliver the male gamete from the pollen grain to the female gamete in the ovule.

Ovule: enclosed within the ovary, and contains the female gamete which has the haploid (n) number of chromosomes.

Fertilisation occurs when the male and female gametes, both of which are haploid, fuse to form a diploid (2n) zygote. At the same time other changes occur which convert some of the ovule cells into food reserves.

Male gamete: has the haploid (n) number of chromo-somes and was formed by meiosis (reduction division) in the anther.

Keys and classification

* the FIVE KINGDOMS

A key enables identification of an organism by observation of its characteristics. Close observation allows a series of questions (the branch points in this key) to be answered, eventually leading to the organism being studied.

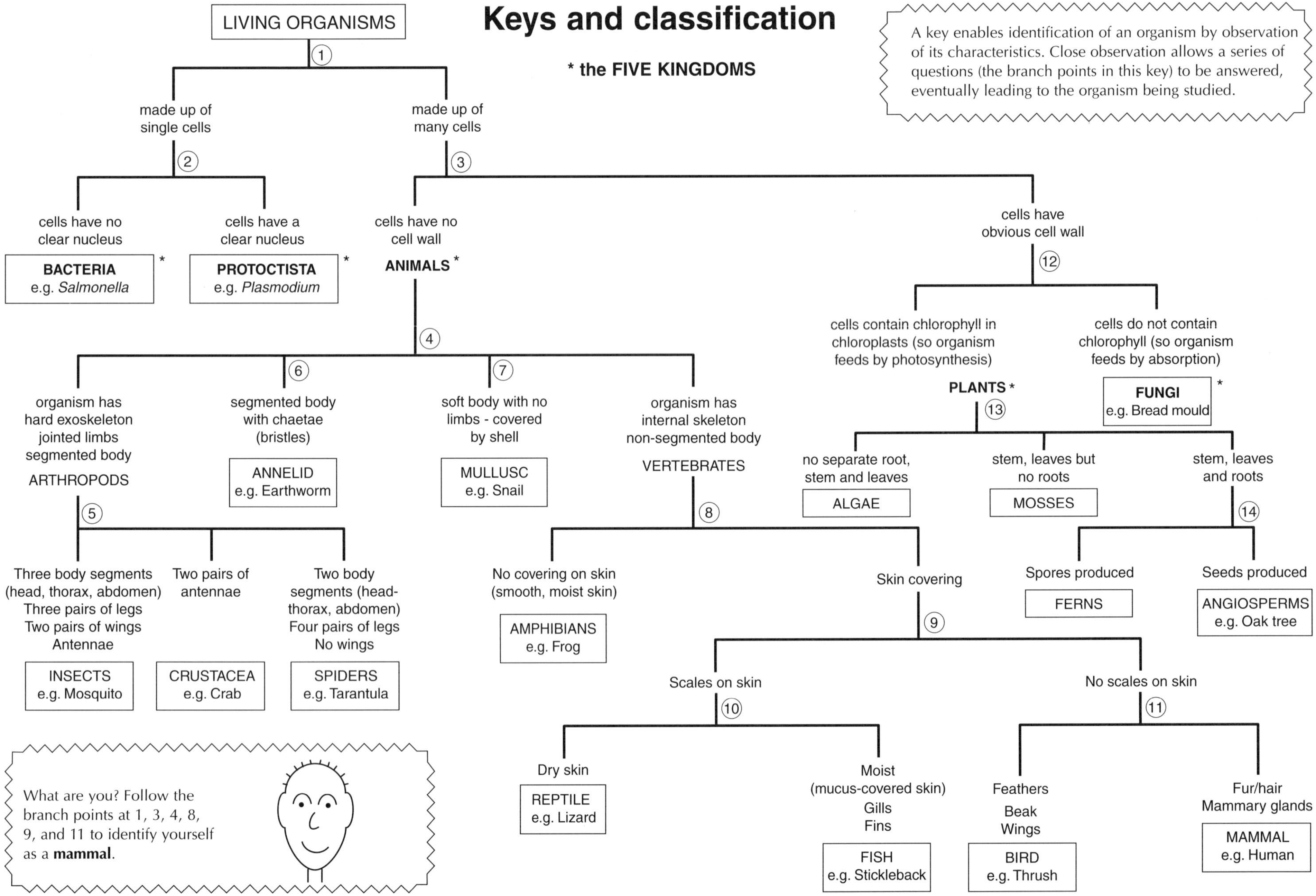

- ① LIVING ORGANISMS
  - made up of single cells
    - ② cells have no clear nucleus → **BACTERIA*** e.g. *Salmonella*
    - cells have a clear nucleus → **PROTOCTISTA*** e.g. *Plasmodium*
  - ③ made up of many cells
    - ④ cells have no cell wall → **ANIMALS***
      - ⑦ soft body with no limbs - covered by shell → **MULLUSC** e.g. Snail
      - ⑥ segmented body with chaetae (bristles) → **ANNELID** e.g. Earthworm
      - ⑤ organism has hard exoskeleton, jointed limbs, segmented body → **ARTHROPODS**
        - Three body segments (head, thorax, abdomen), Three pairs of legs, Two pairs of wings, Antennae → **INSECTS** e.g. Mosquito
        - Two pairs of antennae → **CRUSTACEA** e.g. Crab
        - Two body segments (head-thorax, abdomen), Four pairs of legs, No wings → **SPIDERS** e.g. Tarantula
      - ⑧ organism has internal skeleton, non-segmented body → **VERTEBRATES**
        - No covering on skin (smooth, moist skin) → **AMPHIBIANS** e.g. Frog
        - ⑨ Skin covering
          - ⑩ Scales on skin
            - Moist (mucus-covered skin), Gills, Fins → **FISH** e.g. Stickleback
            - Dry skin → **REPTILE** e.g. Lizard
          - ⑪ No scales on skin
            - Feathers, Beak, Wings → **BIRD** e.g. Thrush
            - Fur/hair, Mammary glands → **MAMMAL** e.g. Human
    - ⑫ cells have obvious cell wall
      - cells do not contain chlorophyll (so organism feeds by absorption) → **FUNGI*** e.g. Bread mould
      - cells contain chlorophyll in chloroplasts (so organism feeds by photosynthesis) → **PLANTS***
        - ⑬ no separate root, stem and leaves → **ALGAE**
        - stem, leaves but no roots → **MOSSES**
        - ⑭ stem, leaves and roots
          - Spores produced → **FERNS**
          - Seeds produced → **ANGIOSPERMS** e.g. Oak tree

What are you? Follow the branch points at 1, 3, 4, 8, 9, and 11 to identify yourself as a **mammal**.

Naming and classifying living organisms

THE BINOMIAL SYSTEM OF NOMENCLATURE

- provided by Linnaeus

- successful since
  (a) each species has a unique name
  (b) shows which species are closely related e.g. *Panthera leo* and *Panthera tigris*

- incorporates both species and genus into specific name
  e.g. *Homo sapiens*
  (i) Underline or *italicise*
  (ii) Capital letter for genus, lower case for species.

- the 'naming' of organisms (usually after they have been placed into 'groups' or taxons)

- needed to aid communication between groups of scientists so 'name' must be unambiguous and easily understood

- usually in Latin since
  (a) original scientific language
  (b) universally accepted

HIERARCHICAL CLASSIFICATION OF THE LION

On moving down the hierarchy of groups, note that there are …

MORE SIMILARITIES and FEWER DIFFERENCES between the members

- KINGDOM (Animalia)
  - PHYLUM (Chordata) — Other phyla
    - CLASS (Mammalia) — Other classes
      - ORDER (Carnivora) — Other orders
        - FAMILY (Felidae) — Other families
          - GENUS (*Panthera*) — Other genera
            - SPECIES *Panthera leo* — Other species

All animals are ingestive heterotrophs

All chordates have a notochord (→backbone)

All mammals have fur and mammary glands

All carnivores have well-developed carnassial (flesh-cutting) teeth

All felidae have retractable claws

All *Panthera* (big cats) can roar but cannot purr

All lions can mate and produce fertile offspring with other lions

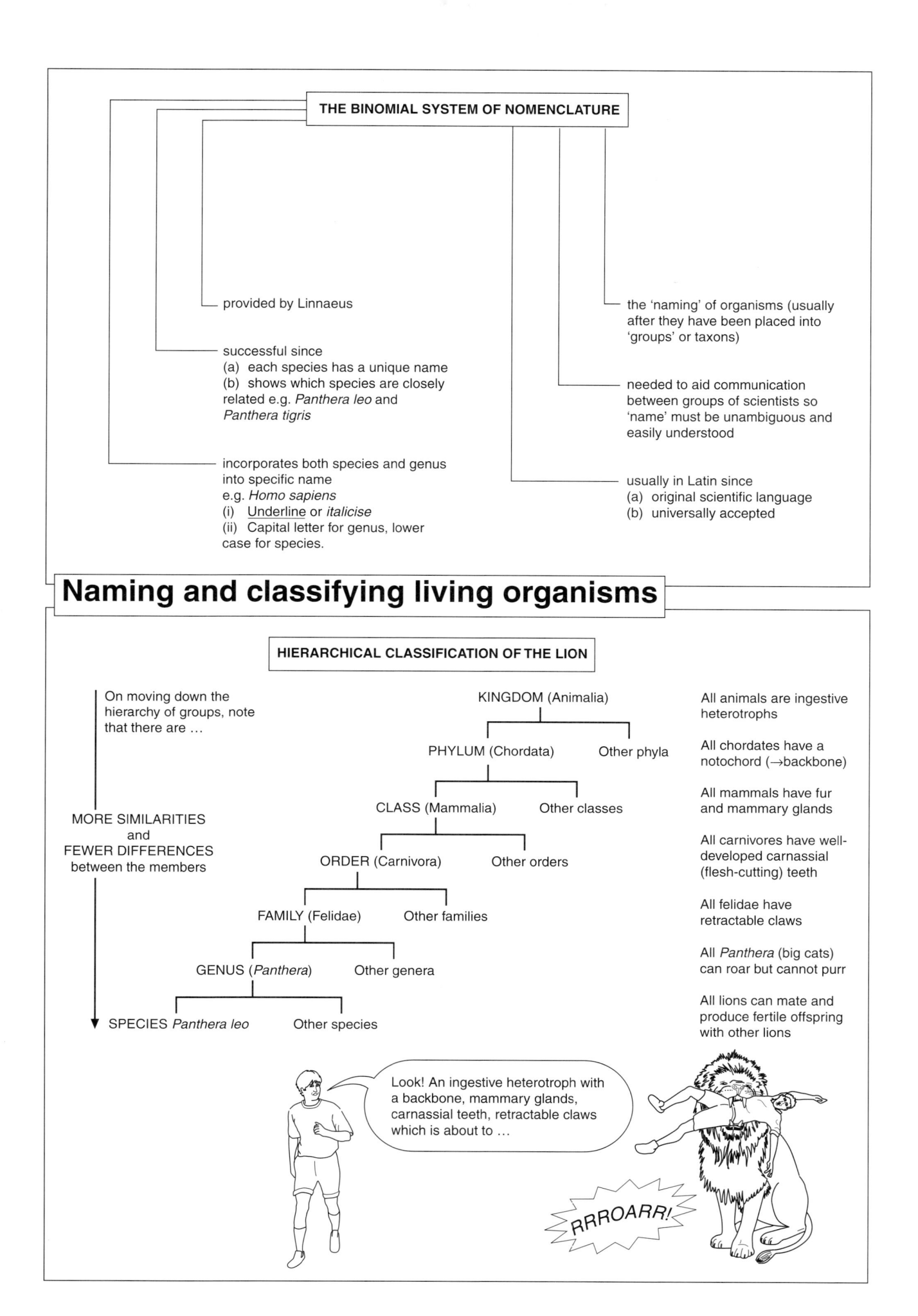

Look! An ingestive heterotroph with a backbone, mammary glands, carnassial teeth, retractable claws which is about to …

RRROARR!

Naming and classifying living organisms 41

Cell division and the human life cycle

HUMAN LIFE CYCLE

Male parent (2n=46) ──── MEIOSIS ──→ sperm (n=23)
                                    → sperm (n=23)

Female parent (2n=46) ──── MEIOSIS ──→ egg (n=23)
                                     → egg (n=23)

Fertilisation - return to normal (diploid) number of chromosomes → zygote (2n=46) → New adult (2n=46)

Mitosis - provides more cells as 'building blocks' of new organism

Meiosis is necessary to halve the chromosome number from diploid (2n) to haploid (n).

All cells contain two sets of chromosomes e.g. humans 2x23 = 46. These chromosomes are homologous i.e. each carries the same genes, though they may be different alleles (alternatives of the same gene).

(for simplicity, only one pair of chromosomes is shown)

MITOSIS: COPYING DIVISION

DNA is exactly replicated

Each chromosome now becomes two identical chromatids joined at centromere.

Chromosomes become attached to a spindle, fibres which run from one pole to the other.

The individual chromosomes line up at the equator (midline) of the cell.

Centromere divides and spindle fibres shorten so that each chromosome becomes two chromatids.

Cell membrane 'pinches in' to separate the two sets of chromatids into two cells.

The original cell has now become two daughter cells
- identical to one another
- identical to parent cell
- same number of chromosomes as parent cell

MEIOSIS: REDUCTION DIVISION

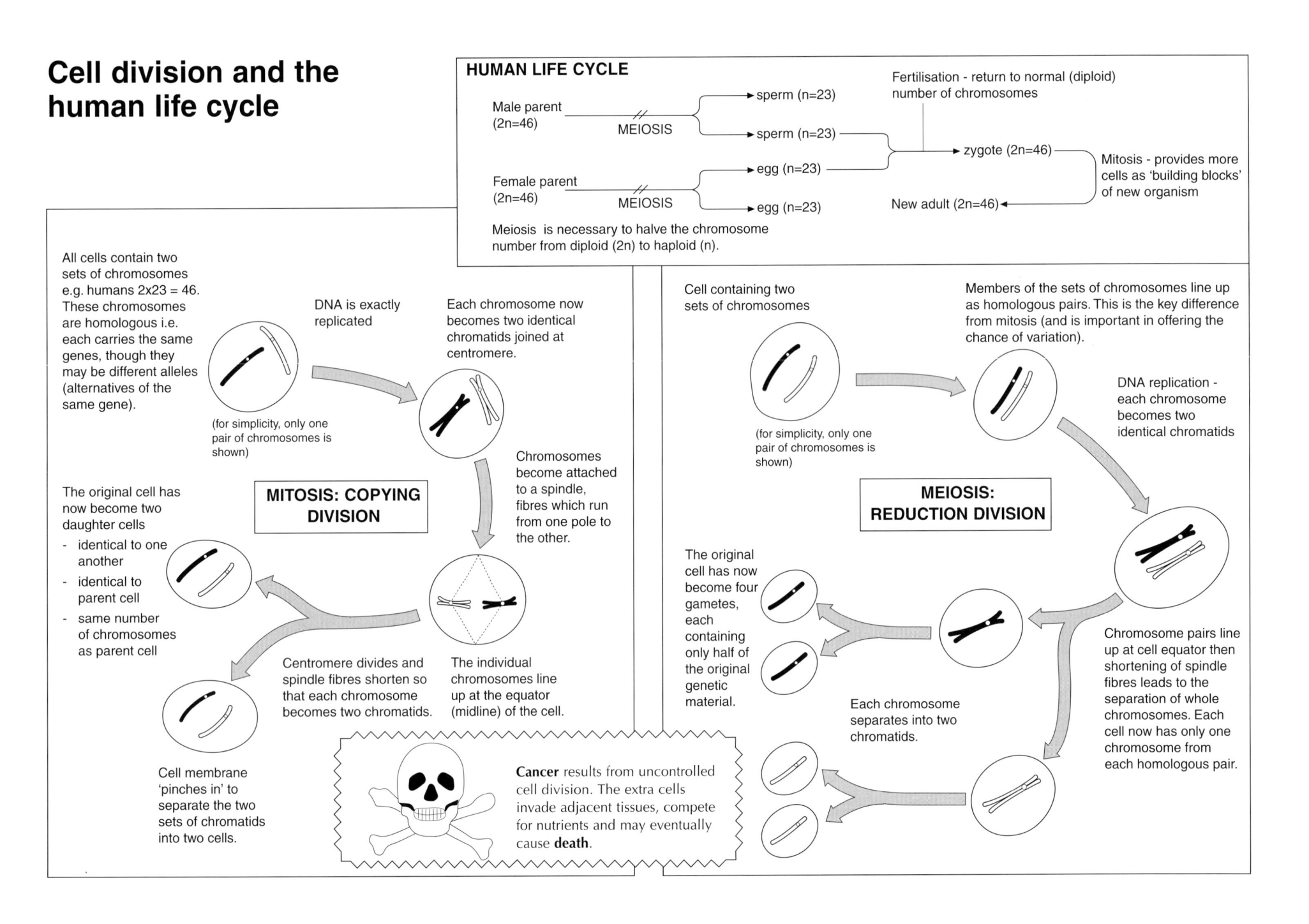

Cell containing two sets of chromosomes

(for simplicity, only one pair of chromosomes is shown)

Members of the sets of chromosomes line up as homologous pairs. This is the key difference from mitosis (and is important in offering the chance of variation).

DNA replication - each chromosome becomes two identical chromatids

Chromosome pairs line up at cell equator then shortening of spindle fibres leads to the separation of whole chromosomes. Each cell now has only one chromosome from each homologous pair.

Each chromosome separates into two chromatids.

The original cell has now become four gametes, each containing only half of the original genetic material.

Cancer results from uncontrolled cell division. The extra cells invade adjacent tissues, compete for nutrients and may eventually cause **death**.

DNA and chromosomes

The nucleus of each human cell (mature red blood cells excepted) contains 46 chromosomes arranged as **23 homologous pairs**.

Homologous pairs of chromosomes carry the same genes (e.g. the gene for eye colour) but may carry alternative forms, or **alleles** of the same gene (e.g. blue on one chromosome, brown on the other)

The complete set of homologous chromosomes carries all of the genetic material for the individual = **genotype**.

This genotype is largely responsible for all of the physical measurable characteristics of this individual (for example, hair and eye colour) = **phenotype**.

GENOTYPE → (EFFECTS OF ENVIRONMENT e.g. food availability amount of sunlight inhaled smoke) → PHENOTYPE

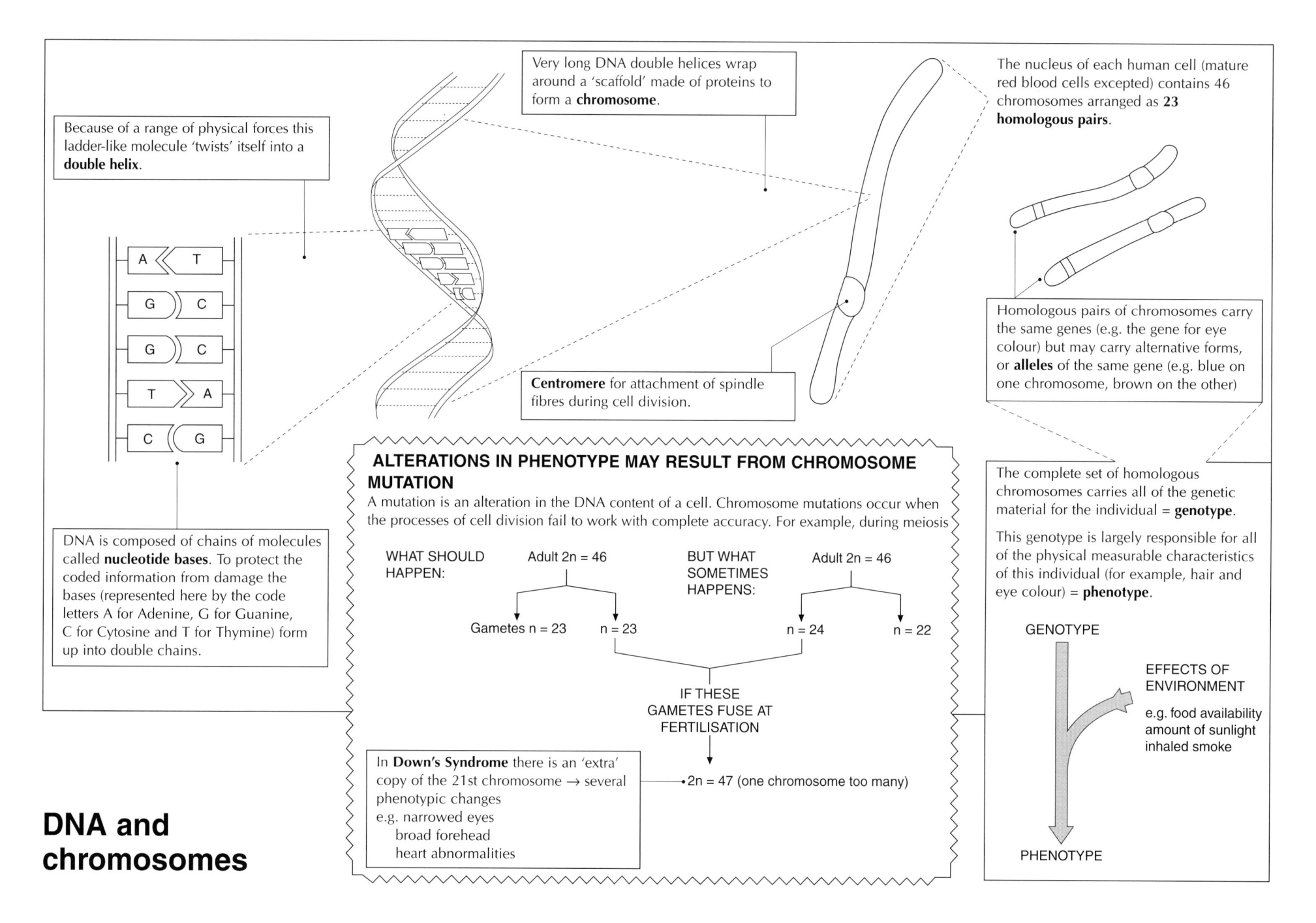

Very long DNA double helices wrap around a 'scaffold' made of proteins to form a **chromosome**.

Centromere for attachment of spindle fibres during cell division.

Because of a range of physical forces this ladder-like molecule 'twists' itself into a **double helix**.

DNA is composed of chains of molecules called **nucleotide bases**. To protect the coded information from damage the bases (represented here by the code letters A for Adenine, G for Guanine, C for Cytosine and T for Thymine) form up into double chains.

A–T
G–C
G–C
T–A
C–G

ALTERATIONS IN PHENOTYPE MAY RESULT FROM CHROMOSOME MUTATION

A mutation is an alteration in the DNA content of a cell. Chromosome mutations occur when the processes of cell division fail to work with complete accuracy. For example, during meiosis

WHAT SHOULD HAPPEN:
Adult 2n = 46 → Gametes n = 23

BUT WHAT SOMETIMES HAPPENS:
Adult 2n = 46 → n = 24 / n = 22

IF THESE GAMETES FUSE AT FERTILISATION → 2n = 47 (one chromosome too many)

In **Down's Syndrome** there is an 'extra' copy of the 21st chromosome → several phenotypic changes
e.g. narrowed eyes
broad forehead
heart abnormalities

DNA, genes and proteins

In the DNA molecule each 'code letter' (nucleotide base) **always** pairs with a specific partner i.e.
A········T
and G········C

```
A---T
G---C
G---C
T---A
C---G
T---A
A---T
C---G
```

This is of great significance in explaining how the DNA molecule is copied during cell division.

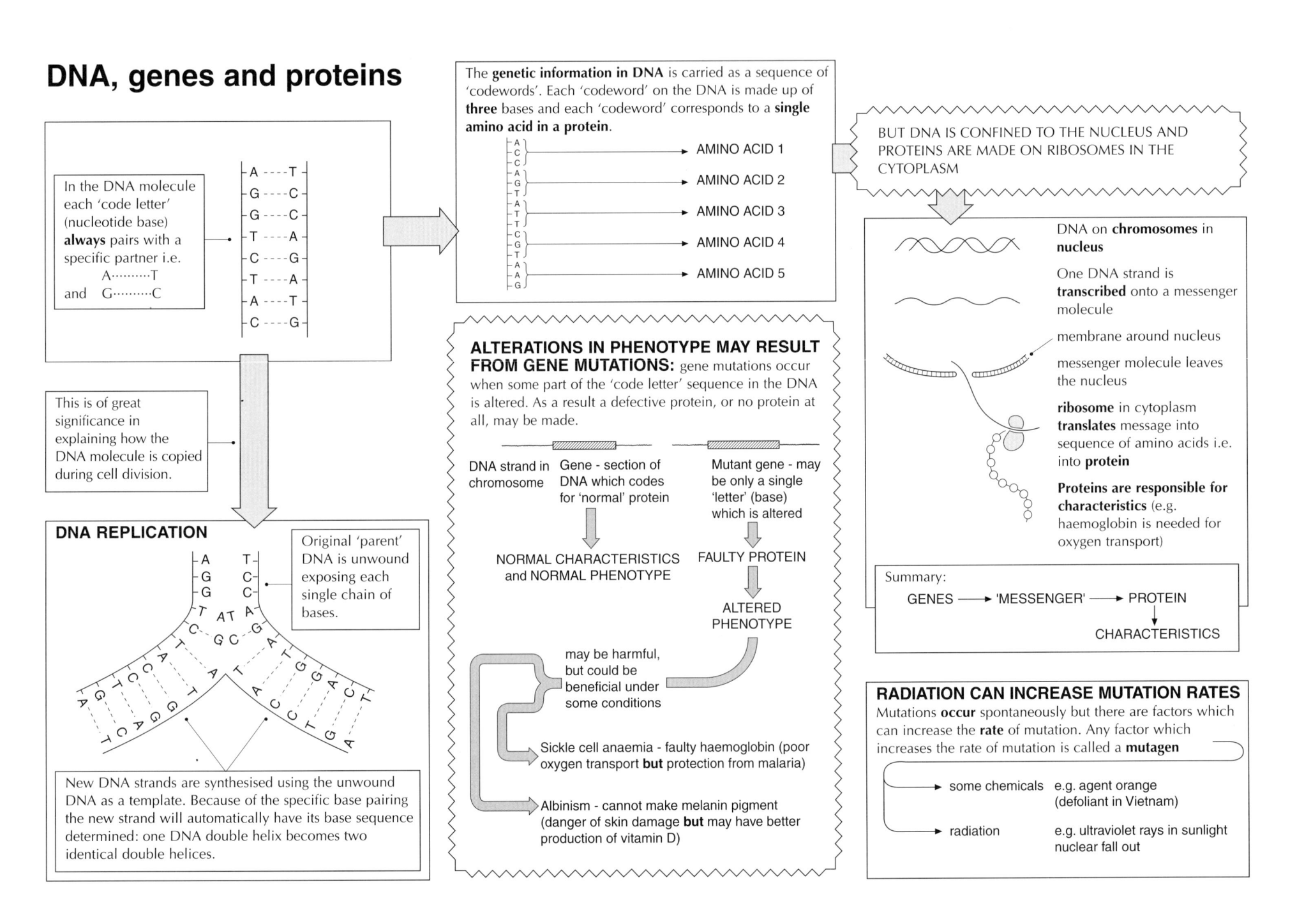

DNA REPLICATION

Original 'parent' DNA is unwound exposing each single chain of bases.

New DNA strands are synthesised using the unwound DNA as a template. Because of the specific base pairing the new strand will automatically have its base sequence determined: one DNA double helix becomes two identical double helices.

The **genetic information in DNA** is carried as a sequence of 'codewords'. Each 'codeword' on the DNA is made up of **three** bases and each 'codeword' corresponds to a **single amino acid in a protein**.

```
A C C → AMINO ACID 1
A C C → AMINO ACID 2
G T T → AMINO ACID 3
A T T → AMINO ACID 4
T C G → AMINO ACID 5
C G A → 
T A T →
A A G →
```

BUT DNA IS CONFINED TO THE NUCLEUS AND PROTEINS ARE MADE ON RIBOSOMES IN THE CYTOPLASM

DNA on **chromosomes** in **nucleus**

One DNA strand is **transcribed** onto a messenger molecule

membrane around nucleus

messenger molecule leaves the nucleus

ribosome in cytoplasm **translates** message into sequence of amino acids i.e. into **protein**

Proteins are responsible for characteristics (e.g. haemoglobin is needed for oxygen transport)

Summary: GENES → 'MESSENGER' → PROTEIN → CHARACTERISTICS

ALTERATIONS IN PHENOTYPE MAY RESULT FROM GENE MUTATIONS:
gene mutations occur when some part of the 'code letter' sequence in the DNA is altered. As a result a defective protein, or no protein at all, may be made.

DNA strand in chromosome

Gene - section of DNA which codes for 'normal' protein

NORMAL CHARACTERISTICS and NORMAL PHENOTYPE

Mutant gene - may be only a single 'letter' (base) which is altered

FAULTY PROTEIN

ALTERED PHENOTYPE

may be harmful, but could be beneficial under some conditions

Sickle cell anaemia - faulty haemoglobin (poor oxygen transport **but** protection from malaria)

Albinism - cannot make melanin pigment (danger of skin damage **but** may have better production of vitamin D)

RADIATION CAN INCREASE MUTATION RATES

Mutations **occur** spontaneously but there are factors which can increase the **rate** of mutation. Any factor which increases the rate of mutation is called a **mutagen**

some chemicals e.g. agent orange (defoliant in Vietnam)

radiation e.g. ultraviolet rays in sunlight nuclear fall out

Cystic fibrosis in humans is an example of monohybrid inheritance

Cystic fibrosis results from an imbalance of chloride ions across the membranes of cells lining some of the major passageways of the body, causing dangerous accumulation of mucus.

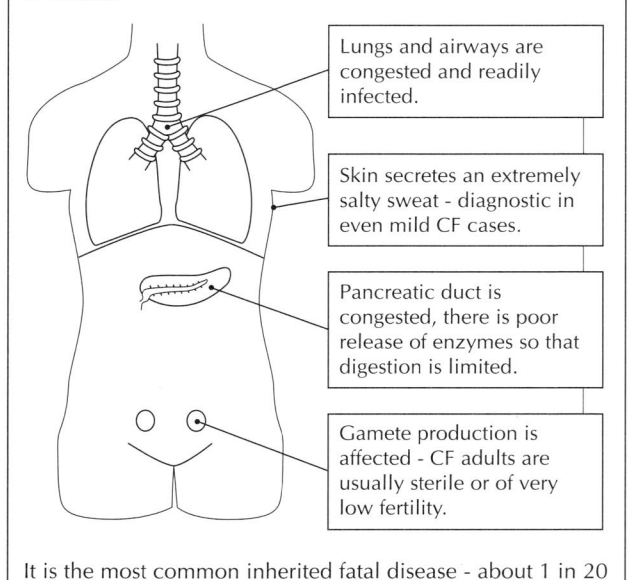

- Lungs and airways are congested and readily infected.
- Skin secretes an extremely salty sweat - diagnostic in even mild CF cases.
- Pancreatic duct is congested, there is poor release of enzymes so that digestion is limited.
- Gamete production is affected - CF adults are usually sterile or of very low fertility.

It is the most common inherited fatal disease - about 1 in 20 white Europeans is a carrier of the mutant allele.

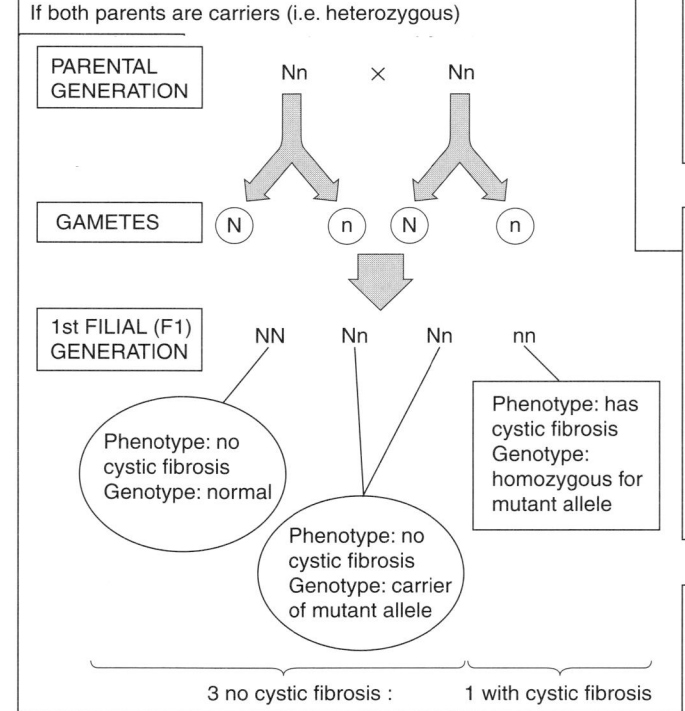

If one parent's genotype is homozygous normal, the other parent's is homozygous mutant, then theoretically:

The homozygous normal parent is represented as NN, the parent with homozygous mutant genotype as nn, since in this case normal is dominant to mutant.

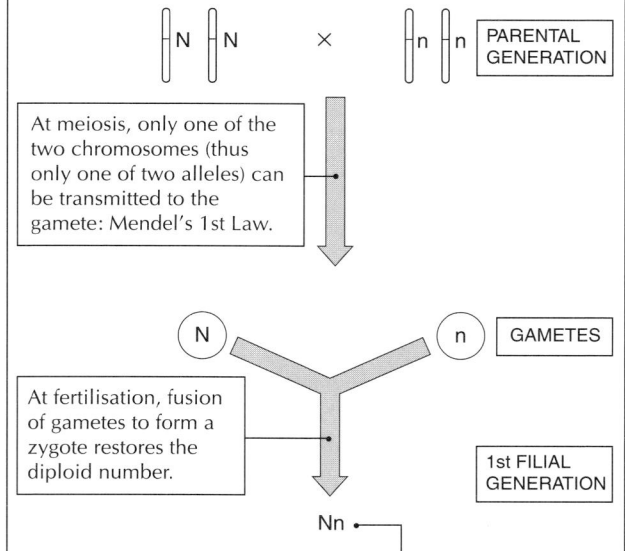

NN × nn — PARENTAL GENERATION

At meiosis, only one of the two chromosomes (thus only one of two alleles) can be transmitted to the gamete: Mendel's 1st Law.

GAMETES: N, n

At fertilisation, fusion of gametes to form a zygote restores the diploid number.

Nn — 1st FILIAL GENERATION

The allele which 'shows up' in this heterozygote is DOMINANT, the other allele (which remains 'hidden') is RECESSIVE.

This individual is **genotypically** heterozygous, but **phenotypically** normal e.g. is a **carrier** of the mutant allele for cystic fibrosis.

If both parents are carriers (i.e. heterozygous)

PARENTAL GENERATION: Nn × Nn

GAMETES: N n N n

1st FILIAL (F1) GENERATION: NN Nn Nn nn

- NN — Phenotype: no cystic fibrosis, Genotype: normal
- Nn — Phenotype: no cystic fibrosis, Genotype: carrier of mutant allele
- nn — Phenotype: has cystic fibrosis, Genotype: homozygous for mutant allele

3 no cystic fibrosis : 1 with cystic fibrosis

NB The 3 : 1 ratio is only approximate unless the number of offspring is very large (unlikely in humans), because
1. alleles may not be distributed between viable gametes in equal numbers
2. fusion of gametes is completely random. It is a matter of chance whether one male gamete fuses with a particular female gamete.

A Punnett square can be used to predict the possible combinations of alleles in the zygote

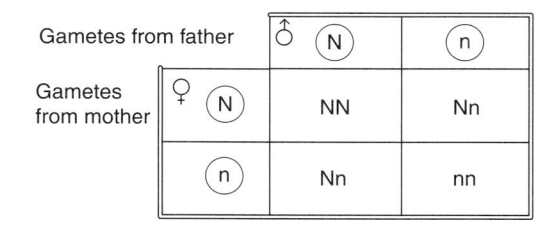

Gametes from mother \ Gametes from father ♂	N	n
♀ N	NN	Nn
n	Nn	nn

Huntington's Chorea is another example of monohybrid inheritance
- Sufferers from this condition usually show no symptoms until middle age (by which time they may have already passed on the mutant allele to their offspring) but then show a rapid, progressive deterioration of the nervous system and loss of co-ordination.
- The mutant allele is **dominant** so that heterozygotes suffer from Huntington's Chorea.
- A cross between two heterozygous parents theoretically produces offspring in the ratio
  3 sufferers : 1 normal
- Individuals with the mutant allele can be detected by sophisticated **gene probes**.

Sex linkage and the inheritance of sex

A KARYOTYPE
is obtained by rearranging photographs of stained chromosomes observed during mitosis. Such a karyotype indicates that

① the chromosomes are arranged in **homologous pairs**. In humans there are 23 pairs and we say that the **diploid number** is 46 ($2n = 46 = 2 \times 23$).

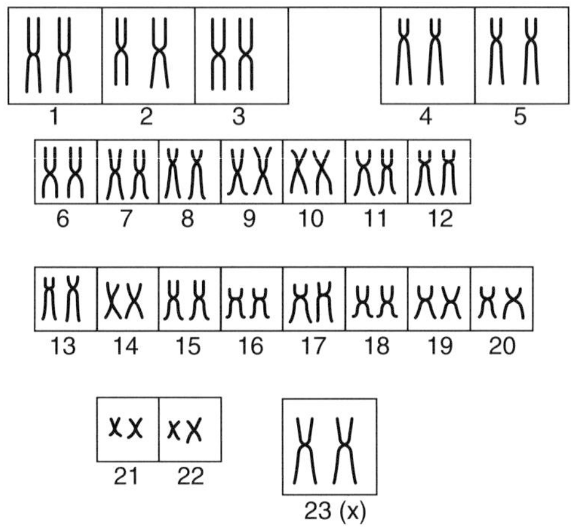

② whereas females have 22 pairs + XX in the karyotype, males have 22 pairs + XY i.e. the 23rd 'pair' would **not** be two copies of the X chromosome.

The Y chromosome is so small that there is little room for any genes other than those responsible for 'maleness', but the X chromosome can carry some genes as well as those for 'femaleness' – these additional genes are **X-linked** (usually described as **sex-linked**).

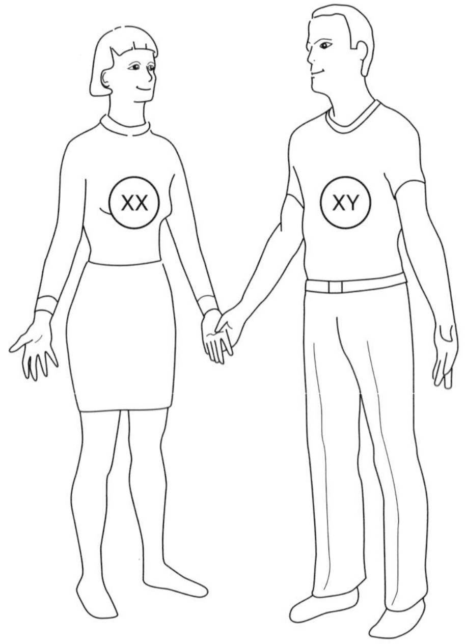

INHERITANCE OF SEX
is a special form of monohybrid inheritance

GAMETES ♂ (XY) → Ⓧ Ⓨ
         ♀ (XX) → Ⓧ Ⓧ

F_1 generation: sex of offspring can be determined from a Punnett square

	♂ GAMETES	
♀ GAMETES	X	Y
X	XX (female)	XY (male)
X	XX (female)	XY (male)

- theoretically there should be 1 : 1 ratio of male : female
- the male gamete determines the sex of the offspring

SEX LINKAGE:
Genes on the X chromosome can be inherited by either sex, but the male can only receive **one** of the possible alleles as he **must** be XY. The female will be XX and thus may be **either homozygous or heterozygous** for the X-linked allele.

For example, **haemophilia** (in which blood fails to clot properly) is an X-linked condition.

Normal allele = H, mutant allele = h

PARENTS $X^H X^h$ × $X^H Y$

          female, normal   male, normal
          phenotype but   phenotype
          a carrier of the
          mutant allele

GAMETES X^H X^h X^H Y

F_1 OFFSPRING

♂ GAMETES →	X^H	Y
♀ GAMETES ↓		
X^H	$X^H X^H$	$X^H Y$
X^h	$X^H X^h$	$X^h Y$

i.e. two 'normal' parents can produce a haemophiliac son

N.B. A haemophiliac daughter ($X^h X^h$) could only be produced if **both** parents contributed X^h i.e. if the father was haemophiliac, in which case his condition would be known and the daughter's condition might be expected.

Another important X-linked condition is **red-green colour blindness**.

Many more males than females cannot distinguish red from green

46 Sex linkage and the inheritance of sex

Variation and natural selection may lead to evolution of species

HOW GENETIC VARIATION MAY ARISE

- both chromosomal and gene **mutation** provide **raw material for variation**
- meiosis and sexual reproduction then re-assort this **raw material** to provide **many new genotypes**.

① **MEIOSIS: CROSSING OVER** and
② **INDEPENDENT ASSORTMENT**

① Homologous chromosomes pair up as meiosis begins

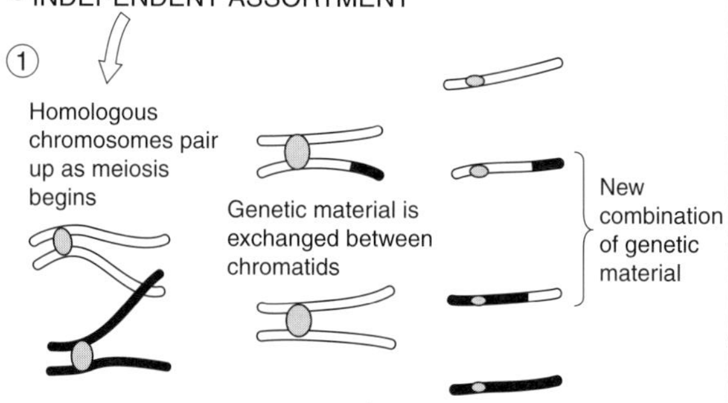

Genetic material is exchanged between chromatids

New combination of genetic material

② When homologous pairs line up in meiosis, they may do so in different ways so that different combinations of alleles will appear in the gametes

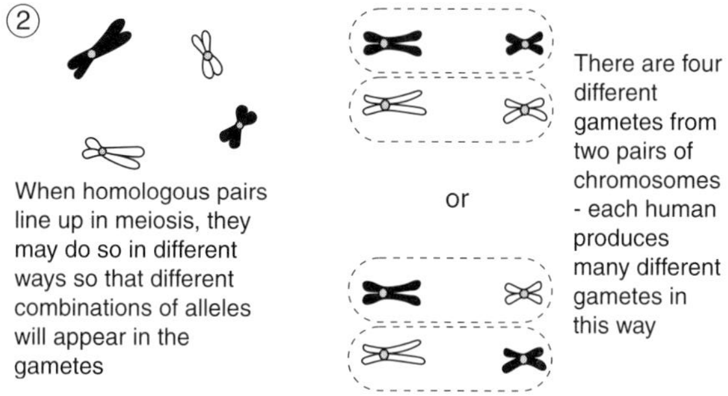

There are four different gametes from two pairs of chromosomes - each human produces many different gametes in this way

③ **FERTILISATION**

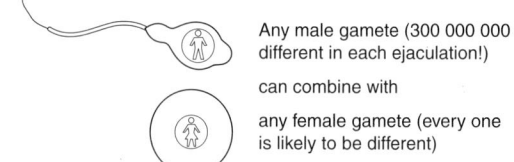

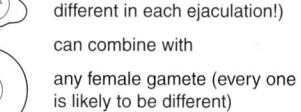

Any male gamete (300 000 000 different in each ejaculation!) can combine with any female gamete (every one is likely to be different)

MUTATION + ① + ② + ③ ⟶ VIRTUALLY LIMITLESS VARIATION

Environmental resistance describes the factors, e.g. food availability, which limit the growth of populations. Animals and plants which are best **adapted** to their environment suffer from less environmental resistance. **Adaptations** are often structural, but they can also be biochemical or behavioural.

These adaptations arise because there is **variation** between the different members of a population. Charles Darwin studied many examples of these adaptations, and published his conclusions in *The Origin of Species by means of Natural Selection*. His observations can be summarised

Over-production: All organisms produce more offspring than can possibly survive, and yet populations remain relatively stable.

e.g. a female peppered moth may lay 500 eggs, but the moth population does not increase by 25000%!

Struggle for existence: Organisms experience environmental resistance i.e. they compete for the limited resources within the environment.

e.g. several moths may try to feed on the same nectar-producing flower.

Variation: Within the population there may be some characteristics which make the organisms which possess them more suited for this severe competition.

e.g. moths might be stronger fliers, have better feeding mouthparts, be better camouflaged while resting or be less affected by rain.

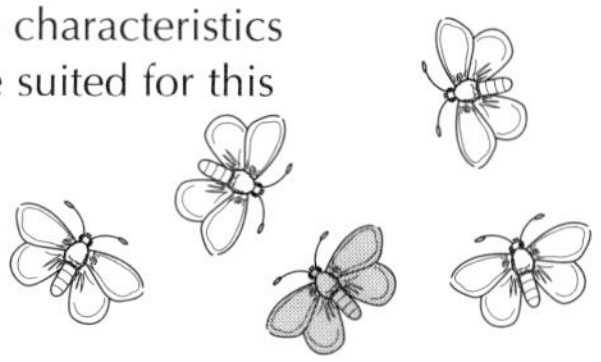

Survival of the fittest: Individuals which are most successful in the struggle for existence (i.e. which are the best suited/adapted to their environment) will survive more easily than those without these advantages.

e.g. peppered moths: dark coloured moths resting on soot-covered tree trunks will be less likely to be captured by predators.

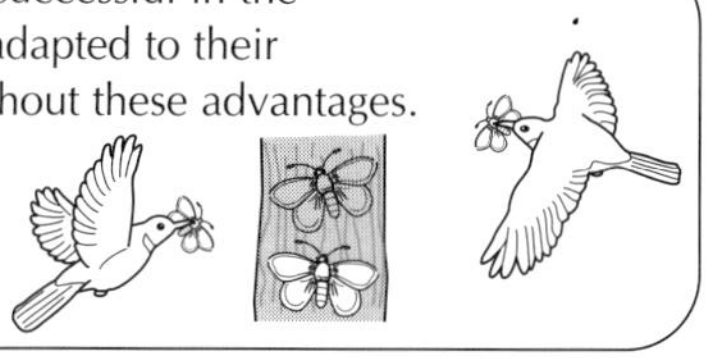

Advantageous characteristics are passed on to offspring: The well-adapted individuals breed more successfully than those which are less well-adapted - they pass on their genes to the next generation. This process is called **natural selection**.

e.g. dark coloured moth parents will produce dark coloured moth offspring.

Summary

VARIATION —NATURAL SELECTION→ ADAPTATION

↓ CONTINUED NATURAL SELECTION

HIGHLY-ADAPTED POPULATIONS = NEW SPECIES

Variation, natural selection and evolution 47

Genetic engineering (recombinant DNA technology)

depends on enzymes and culture of microorganisms

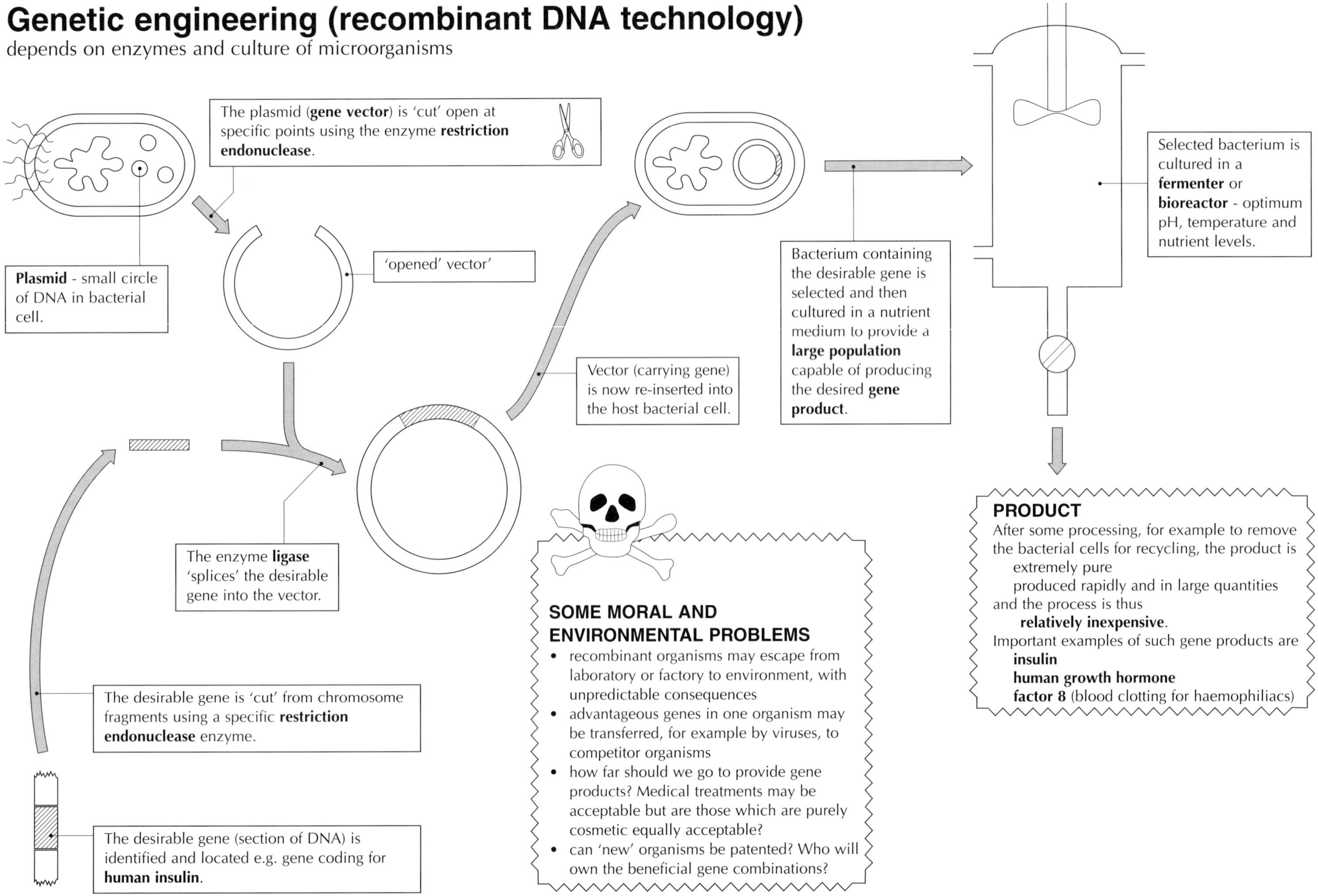

Selected bacterium is cultured in a fermenter or bioreactor - optimum pH, temperature and nutrient levels.

Bacterium containing the desirable gene is selected and then cultured in a nutrient medium to provide a large population capable of producing the desired gene product.

Vector (carrying gene) is now re-inserted into the host bacterial cell.

The plasmid (**gene vector**) is 'cut' open at specific points using the enzyme **restriction endonuclease**.

'opened' vector

Plasmid - small circle of DNA in bacterial cell.

The enzyme **ligase** 'splices' the desirable gene into the vector.

The desirable gene is 'cut' from chromosome fragments using a specific **restriction endonuclease** enzyme.

The desirable gene (section of DNA) is identified and located e.g. gene coding for **human insulin**.

PRODUCT

After some processing, for example to remove the bacterial cells for recycling, the product is extremely pure produced rapidly and in large quantities and the process is thus **relatively inexpensive**. Important examples of such gene products are
insulin
human growth hormone
factor 8 (blood clotting for haemophiliacs)

SOME MORAL AND ENVIRONMENTAL PROBLEMS

- recombinant organisms may escape from laboratory or factory to environment, with unpredictable consequences
- advantageous genes in one organism may be transferred, for example by viruses, to competitor organisms
- how far should we go to provide gene products? Medical treatments may be acceptable but are those which are purely cosmetic equally acceptable?
- can 'new' organisms be patented? Who will own the beneficial gene combinations?

48 Genetic engineering (recombinant DNA technology)

Gene transfer can promote desirable characteristics

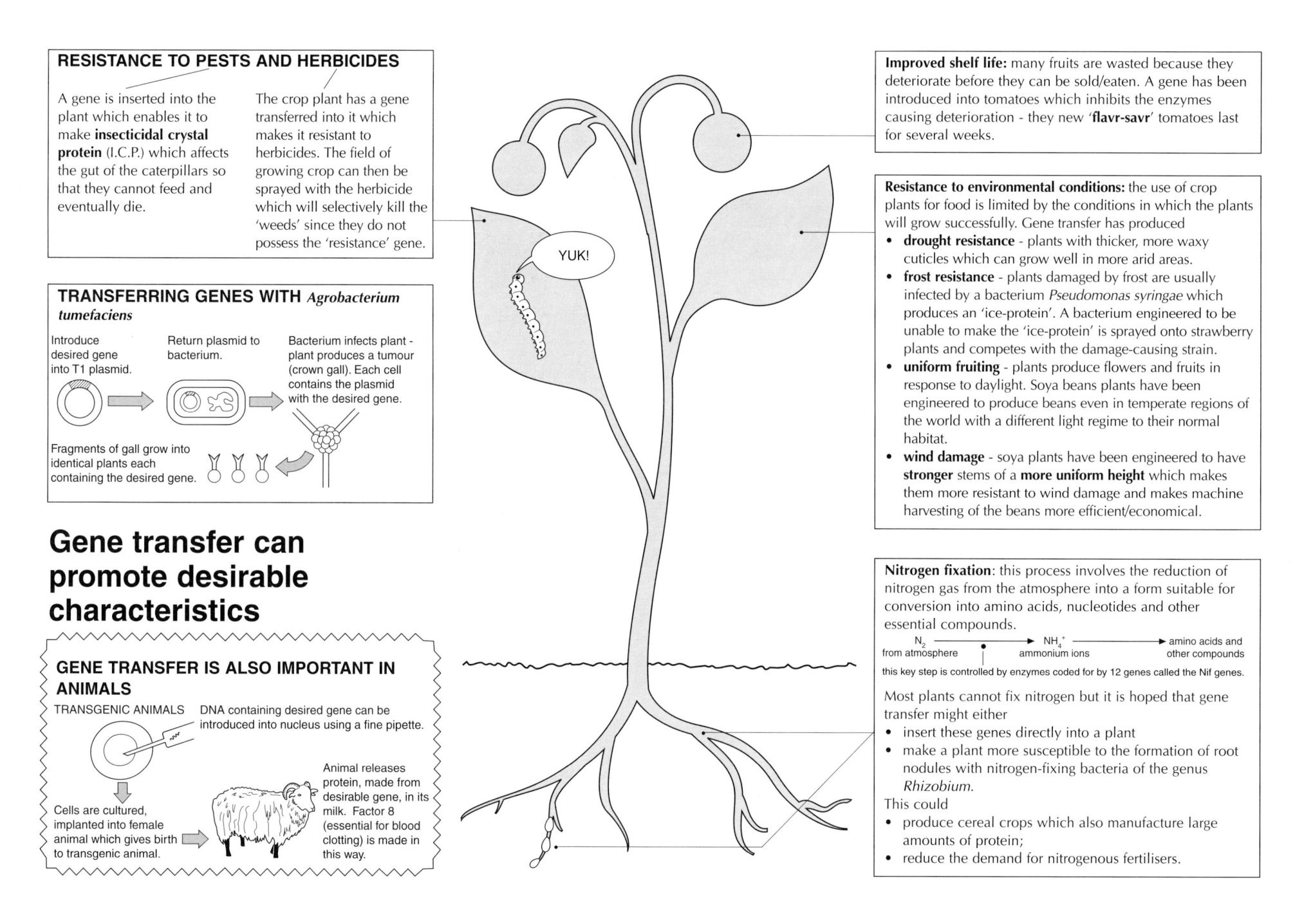

Improved shelf life: many fruits are wasted because they deteriorate before they can be sold/eaten. A gene has been introduced into tomatoes which inhibits the enzymes causing deterioration - they new **'flavr-savr'** tomatoes last for several weeks.

Resistance to environmental conditions: the use of crop plants for food is limited by the conditions in which the plants will grow successfully. Gene transfer has produced

- **drought resistance** - plants with thicker, more waxy cuticles which can grow well in more arid areas.
- **frost resistance** - plants damaged by frost are usually infected by a bacterium *Pseudomonas syringae* which produces an 'ice-protein'. A bacterium engineered to be unable to make the 'ice-protein' is sprayed onto strawberry plants and competes with the damage-causing strain.
- **uniform fruiting** - plants produce flowers and fruits in response to daylight. Soya beans plants have been engineered to produce beans even in temperate regions of the world with a different light regime to their normal habitat.
- **wind damage** - soya plants have been engineered to have **stronger** stems of a **more uniform height** which makes them more resistant to wind damage and makes machine harvesting of the beans more efficient/economical.

Nitrogen fixation: this process involves the reduction of nitrogen gas from the atmosphere into a form suitable for conversion into amino acids, nucleotides and other essential compounds.

$$N_2 \text{ from atmosphere} \longrightarrow NH_4^+ \text{ ammonium ions} \longrightarrow \text{amino acids and other compounds}$$

this key step is controlled by enzymes coded for by 12 genes called the Nif genes.

Most plants cannot fix nitrogen but it is hoped that gene transfer might either

- insert these genes directly into a plant
- make a plant more susceptible to the formation of root nodules with nitrogen-fixing bacteria of the genus *Rhizobium*.

This could

- produce cereal crops which also manufacture large amounts of protein;
- reduce the demand for nitrogenous fertilisers.

RESISTANCE TO PESTS AND HERBICIDES

A gene is inserted into the plant which enables it to make **insecticidal crystal protein** (I.C.P.) which affects the gut of the caterpillars so that they cannot feed and eventually die.

The crop plant has a gene transferred into it which makes it resistant to herbicides. The field of growing crop can then be sprayed with the herbicide which will selectively kill the 'weeds' since they do not possess the 'resistance' gene.

TRANSFERRING GENES WITH *Agrobacterium tumefaciens*

Introduce desired gene into Ti plasmid.

Return plasmid to bacterium.

Bacterium infects plant - plant produces a tumour (crown gall). Each cell contains the plasmid with the desired gene.

Fragments of gall grow into identical plants each containing the desired gene.

GENE TRANSFER IS ALSO IMPORTANT IN ANIMALS

TRANSGENIC ANIMALS

DNA containing desired gene can be introduced into nucleus using a fine pipette.

Cells are cultured, implanted into female animal which gives birth to transgenic animal.

Animal releases protein, made from desirable gene, in its milk. Factor 8 (essential for blood clotting) is made in this way.

Cloning - the production of genetically identical individuals - has many applications.

'IN VITRO' (TEST TUBE) FERTILISATION

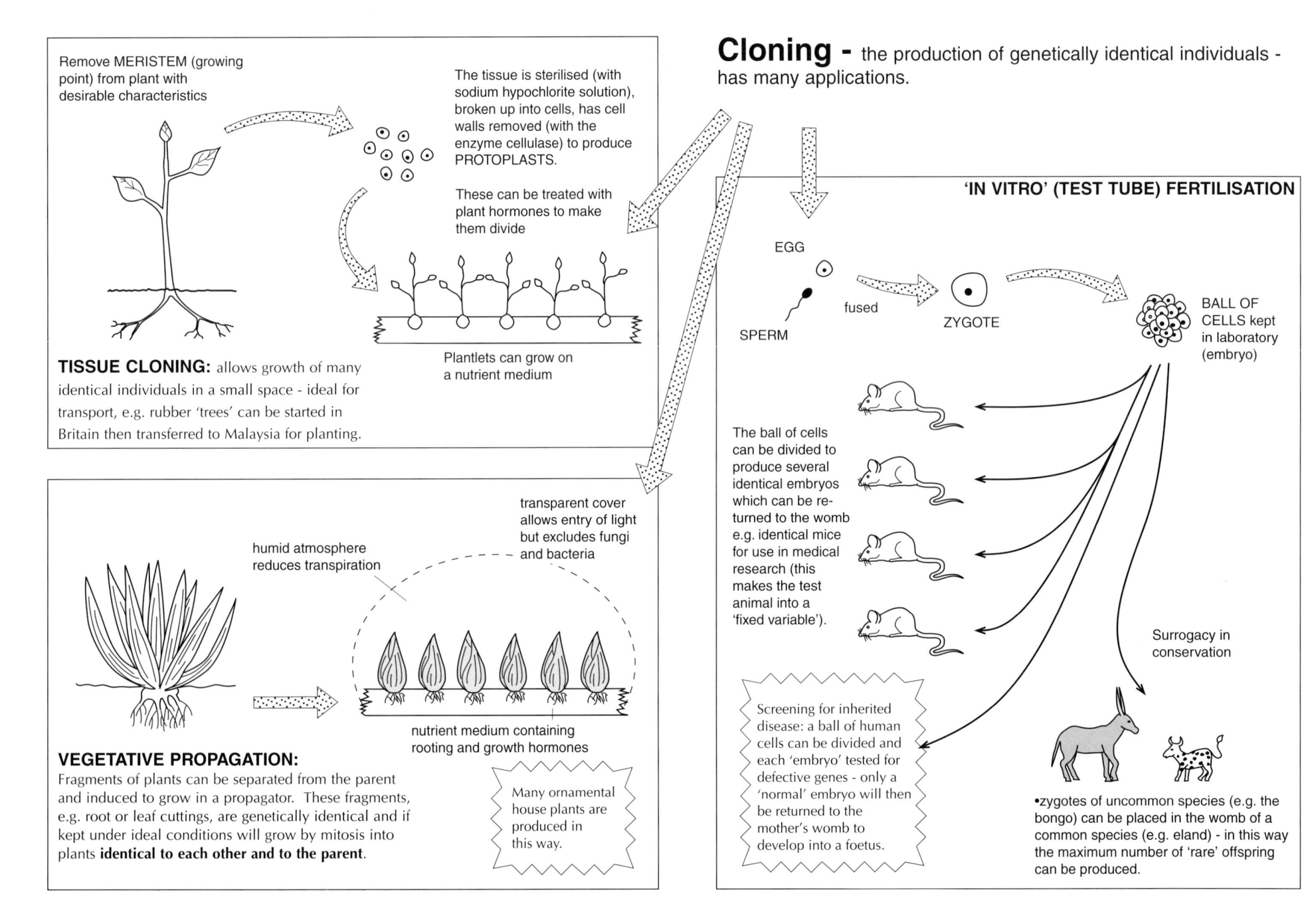

EGG + SPERM → fused → ZYGOTE → BALL OF CELLS kept in laboratory (embryo)

The ball of cells can be divided to produce several identical embryos which can be returned to the womb e.g. identical mice for use in medical research (this makes the test animal into a 'fixed variable').

Screening for inherited disease: a ball of human cells can be divided and each 'embryo' tested for defective genes - only a 'normal' embryo will then be returned to the mother's womb to develop into a foetus.

Surrogacy in conservation
- zygotes of uncommon species (e.g. the bongo) can be placed in the womb of a common species (e.g. eland) - in this way the maximum number of 'rare' offspring can be produced.

TISSUE CLONING
allows growth of many identical individuals in a small space - ideal for transport, e.g. rubber 'trees' can be started in Britain then transferred to Malaysia for planting.

Remove MERISTEM (growing point) from plant with desirable characteristics

The tissue is sterilised (with sodium hypochlorite solution), broken up into cells, has cell walls removed (with the enzyme cellulase) to produce PROTOPLASTS.

These can be treated with plant hormones to make them divide

Plantlets can grow on a nutrient medium

VEGETATIVE PROPAGATION:
Fragments of plants can be separated from the parent and induced to grow in a propagator. These fragments, e.g. root or leaf cuttings, are genetically identical and if kept under ideal conditions will grow by mitosis into plants **identical to each other and to the parent**.

transparent cover allows entry of light but excludes fungi and bacteria

humid atmosphere reduces transpiration

nutrient medium containing rooting and growth hormones

Many ornamental house plants are produced in this way.

50 Cloning

Bioreactors/fermenters exploit microbes for commercial reasons.

VINEGAR: AEROBIC FERMENTATION OF ALCOHOL. Inputs: OXYGEN, MICROBIAL CULTURE, ETHANOL. Output: ETHANOIC ACID (VINEGAR). WOOD SHAVINGS INCREASE SURFACE FOR MICROBIAL FERMENTATION.

BIOGAS: ANAEROBIC FERMENTATION OF HOUSEHOLD WASTE. VENT. METHANE: used as fuel. HOUSEHOLD OR FARMYARD WASTE.

GASOHOL: ANAEROBIC FERMENTATION OF SUGAR CANE WASTE. Inputs: CANE WASTE, YEAST. Output: ALCOHOL: mixed with petrol to provide cheap fuel for internal combustion engine.

WHY USE MICRO-ORGANISMS?
- carry out reactions at moderate temperatures
  → **great savings on fuel**
- are very efficient
  → **less waste and purer products**
- can be 'genetically engineered' to produce **compounds needed for humans**.

Input for microbial culture: the organisms which will carry out the fermentation process are cultured separately until they are growing well.

Nutrient input: the micro-organisms require
- an **energy source** - usually carbohydrate
- **growth materials** - amino acids, or ammonium salts which can be converted to amino acids, are required for protein synthesis.

Constant temperature water jacket: the temperature is controlled so that it is high enough to promote enzyme activity but not so high that enzymes and other proteins in the microbes are denatured.
- this initially involves **heating** to initiate the fermentation
- fermentation releases heat so that later stages may require **cooling**.

Sterile conditions are essential: the culture must be pure and all nutrients/equipment sterile to
- avoid competition for expensive nutrients
- limit the danger of disease-causing organisms contaminating the product.

Paddle stirrers: continuously mix the contents of the bioreactor
- ensure micro-organisms are always in contact with nutrients
- ensures an even temperature throughout the fermentation mixture
- for aerobic (oxygen-requiring) fermentations the mixing may be carried out by an **airstream**.

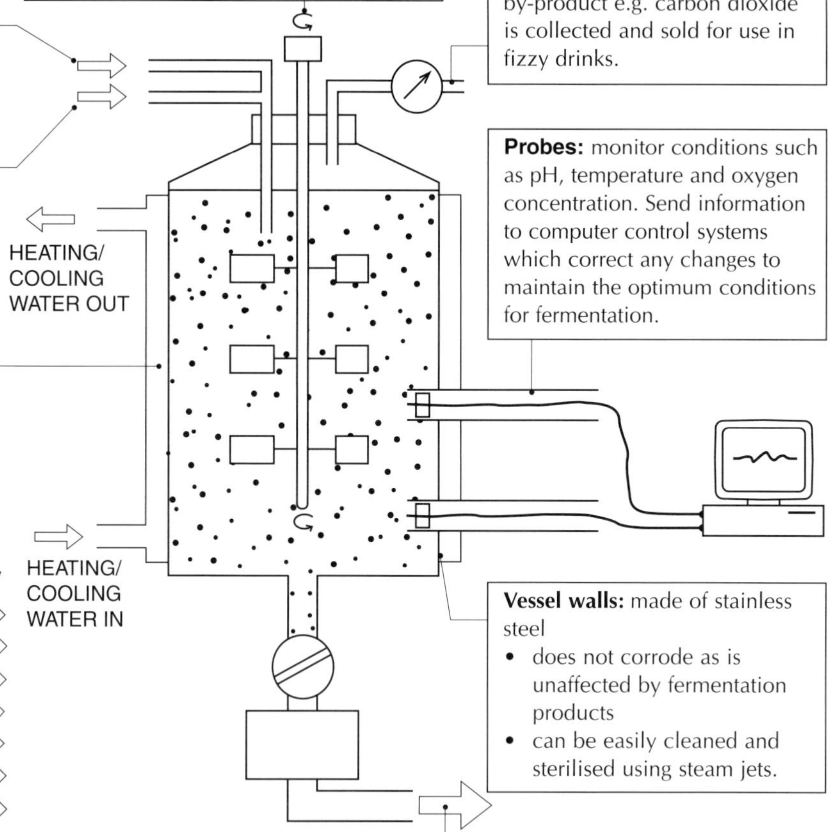

Gas outlet: gas may be evolved during fermentation. This must be released to avoid pressure damage, and may be a valuable by-product e.g. carbon dioxide is collected and sold for use in fizzy drinks.

Probes: monitor conditions such as pH, temperature and oxygen concentration. Send information to computer control systems which correct any changes to maintain the optimum conditions for fermentation.

Vessel walls: made of stainless steel
- does not corrode as is unaffected by fermentation products
- can be easily cleaned and sterilised using steam jets.

HEATING/COOLING WATER OUT. HEATING/COOLING WATER IN.

Further processing of product may be necessary
- to separate the micro-organism from the desired product. In some fermentation systems these micro-organisms may then be returned to the vessel to continue the process.
- to prepare the product for sale or distribution - this often involves **drying** or **crystallisation**.

WASTE NOT, WANT NOT!
- ideally the nutrients should be waste products of other reactions e.g. remains of crops grown for other purposes.
- the micro-organisms should be used to start or maintain new cultures.

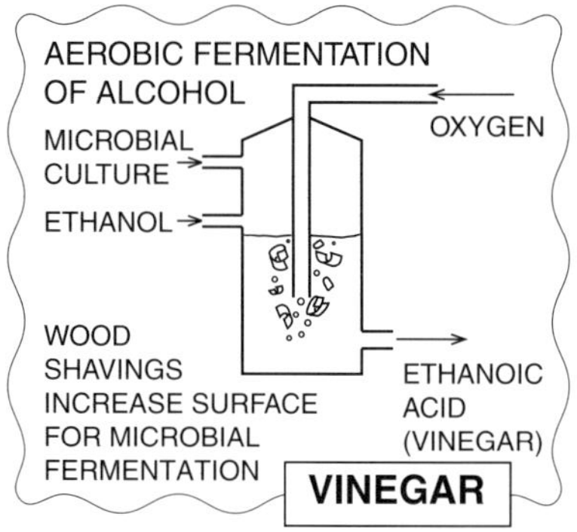

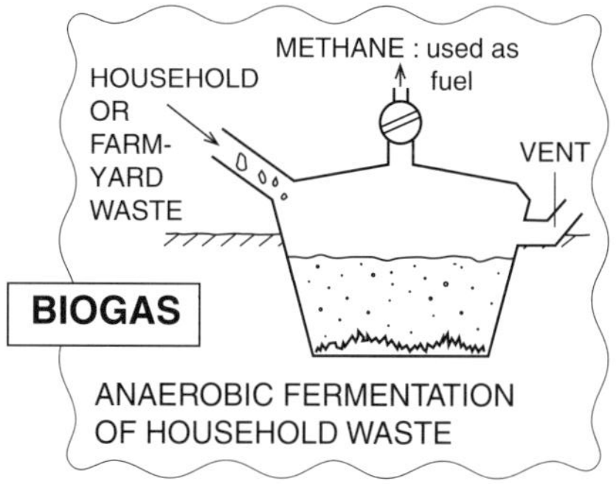

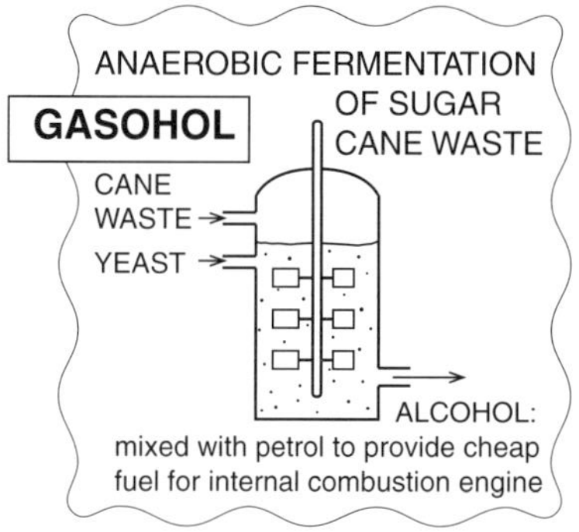

PENICILLIN IS AN ANTIBIOTIC

The *Penicillium* mould may make products which it secretes into its environment to kill off any disease-causing or competitive micro-organisms. **A compound made by a mould to kill off any other micro-organism is called an ANTIBIOTIC.**

Antibiotics are valuable in the control of bacterial diseases because they affect bacterial cells but rarely affect human or animal cells.

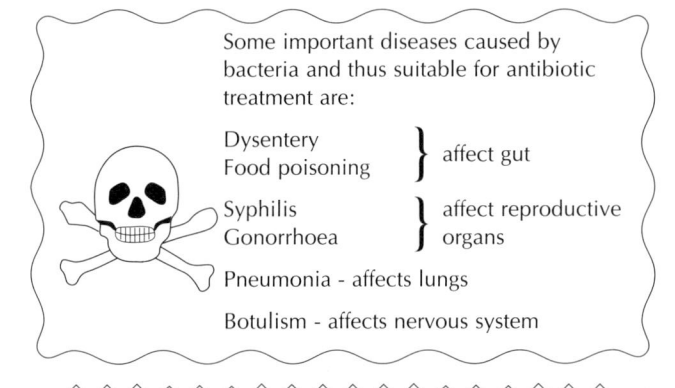

Some important diseases caused by bacteria and thus suitable for antibiotic treatment are:

- Dysentery } affect gut
- Food poisoning
- Syphilis } affect reproductive organs
- Gonorrhoea
- Pneumonia - affects lungs
- Botulism - affects nervous system

RESISTANT BACTERIAL STRAINS MAY DEVELOP

Bacteria breed rapidly and populations are enormous. Within these populations rare **mutations** may produce cells which are **antibiotic-resistant**. Over-use of antibiotics may destroy 'normal' bacterial cells but allow 'resistant' cells to survive. These may multiply to form a resistant population.

One resistant cell in population → ANTIBIOTIC TREATMENT → Only resistant cell survives → CELL DIVISION → Whole population is now resistant

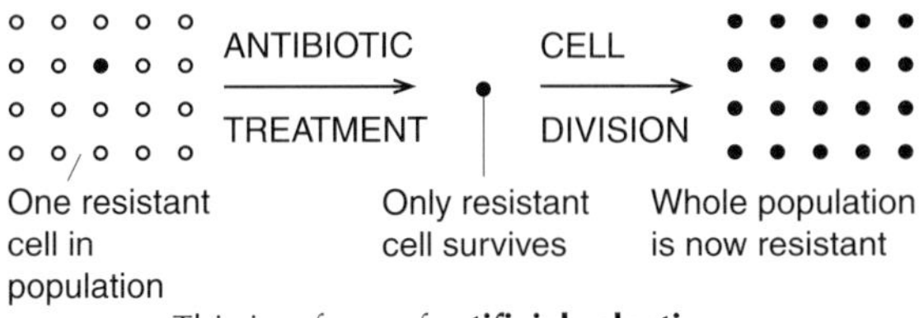

This is a form of **artificial selection**.

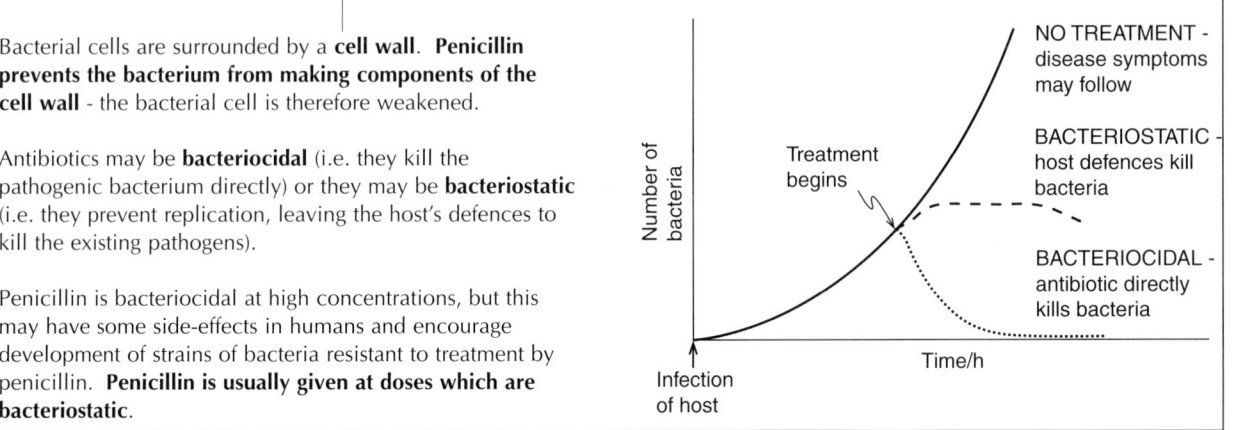

Graph: Number of bacteria vs Time/h, showing Infection of host, Treatment begins, with curves for:
- NO TREATMENT - disease symptoms may follow
- BACTERIOSTATIC - host defences kill bacteria
- BACTERIOCIDAL - antibiotic directly kills bacteria

***PENICILLIUM* MOULD** is a **saprotroph** which reproduces by **sporulation**.

Mould feeds by secreting enzymes onto food and absorbing soluble products. Well-nourished moulds produce vertical hyphae.

Vertical hypha develops spore case containing eight reproductive spores.

Spore case splits and spores are expelled. They are light and thus easily dispersed.

If spore lands on suitable food it germinates and the cycle begins again.

Other, related topics can be found on pages ....

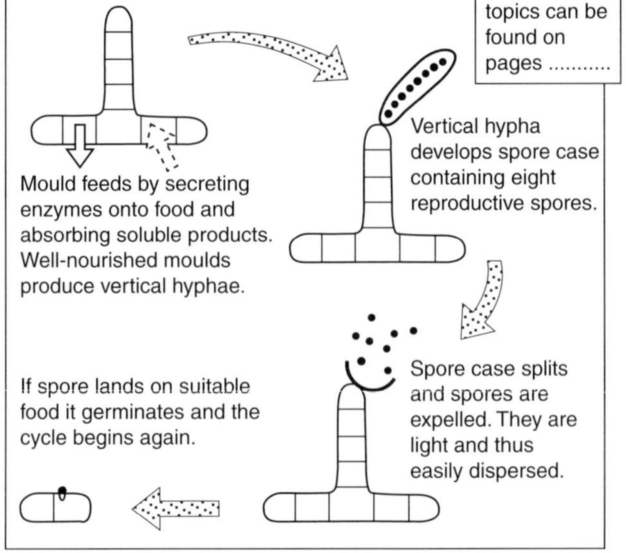

ACTION OF PENICILLIN

Bacteria multiply very rapidly by **binary fission** i.e. they grow and then divide into two. This can happen very rapidly (every 20 minutes), causing disease as the bacterial metabolism competes with or inhibits the activity of human cells.

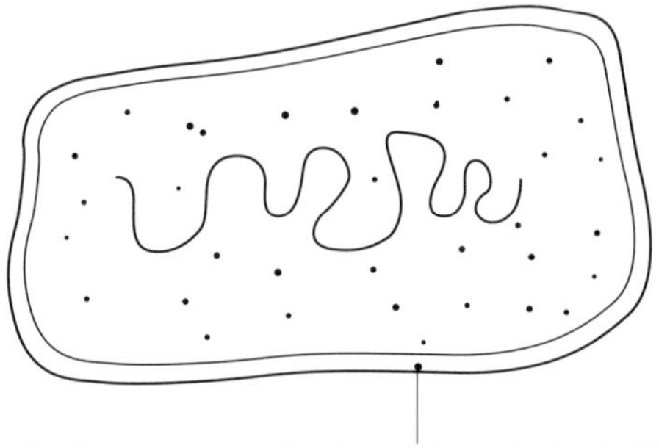

Bacterial cells are surrounded by a **cell wall**. Penicillin **prevents the bacterium from making components of the cell wall** - the bacterial cell is therefore weakened.

Antibiotics may be **bacteriocidal** (i.e. they kill the pathogenic bacterium directly) or they may be **bacteriostatic** (i.e. they prevent replication, leaving the host's defences to kill the existing pathogens).

Penicillin is bacteriocidal at high concentrations, but this may have some side-effects in humans and encourage development of strains of bacteria resistant to treatment by penicillin. **Penicillin is usually given at doses which are bacteriostatic.**

LARGE SCALE PRODUCTION OF PENICILLIN

is a good example of the commercial use of fermentation reactions.

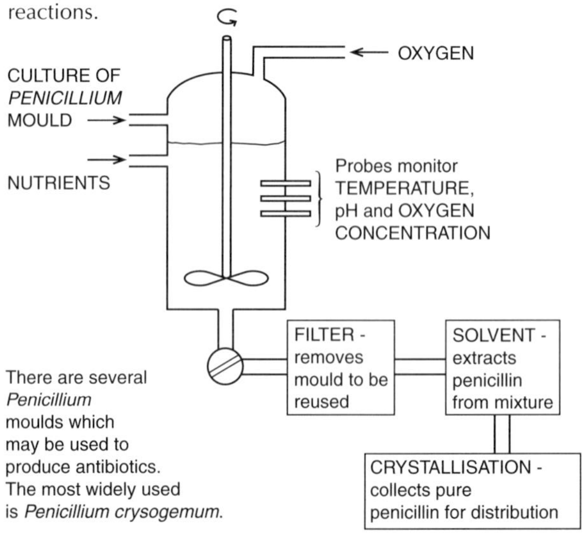

CULTURE OF *PENICILLIUM* MOULD →
NUTRIENTS →
OXYGEN →

Probes monitor TEMPERATURE, pH and OXYGEN CONCENTRATION

FILTER - removes mould to be reused

SOLVENT - extracts penicillin from mixture

CRYSTALLISATION - collects pure penicillin for distribution

There are several *Penicillium* moulds which may be used to produce antibiotics. The most widely used is *Penicillium crysogenum*.

Bacteria and food production

Bacterial cells are surrounded by a cell wall. They have cytoplasm **but no nucleus**. Many of them are **heterotrophic** - they feed on organic foods.

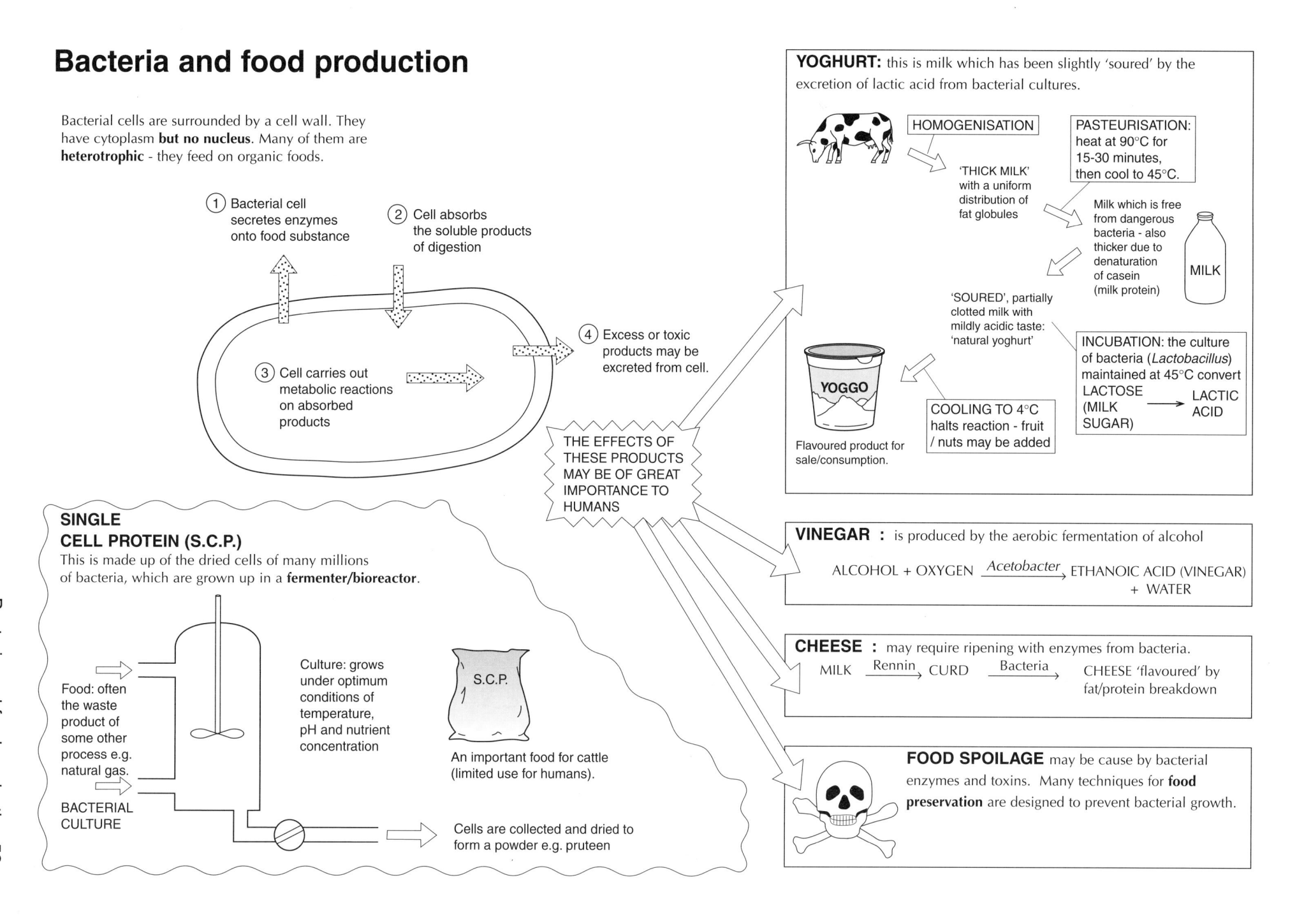

YOGHURT: this is milk which has been slightly 'soured' by the excretion of lactic acid from bacterial cultures.

PASTEURISATION: heat at 90°C for 15-30 minutes, then cool to 45°C.

Milk which is free from dangerous bacteria - also thicker due to denaturation of casein (milk protein)

HOMOGENISATION

'THICK MILK' with a uniform distribution of fat globules

INCUBATION: the culture of bacteria (*Lactobacillus*) maintained at 45°C convert

LACTOSE (MILK SUGAR) → LACTIC ACID

'SOURED', partially clotted milk with mildly acidic taste: 'natural yoghurt'

COOLING TO 4°C halts reaction - fruit / nuts may be added

Flavoured product for sale/consumption.

VINEGAR: is produced by the aerobic fermentation of alcohol

ALCOHOL + OXYGEN $\xrightarrow{Acetobacter}$ ETHANOIC ACID (VINEGAR) + WATER

CHEESE: may require ripening with enzymes from bacteria.

MILK $\xrightarrow{Rennin}$ CURD $\xrightarrow{Bacteria}$ CHEESE 'flavoured' by fat/protein breakdown

FOOD SPOILAGE may be cause by bacterial enzymes and toxins. Many techniques for **food preservation** are designed to prevent bacterial growth.

THE EFFECTS OF THESE PRODUCTS MAY BE OF GREAT IMPORTANCE TO HUMANS

① Bacterial cell secretes enzymes onto food substance

② Cell absorbs the soluble products of digestion

③ Cell carries out metabolic reactions on absorbed products

④ Excess or toxic products may be excreted from cell.

SINGLE CELL PROTEIN (S.C.P.)

This is made up of the dried cells of many millions of bacteria, which are grown up in a **fermenter/bioreactor**.

Food: often the waste product of some other process e.g. natural gas.

BACTERIAL CULTURE

Culture: grows under optimum conditions of temperature, pH and nutrient concentration

Cells are collected and dried to form a powder e.g. pruteen

An important food for cattle (limited use for humans).

Anaerobic respiration

involves incomplete breakdown of glucose.

EXERCISE AND LACTIC ACID

Before exercise, or during mild exercise, the blood can deliver enough oxygen to keep aerobic respiration at a level high enough to supply the body's demands for oxygen.

During severe exercise oxygen delivery increases but not enough to maintain aerobic respiration. The body's demands for energy must now be partially met by anaerobic respiration. Lactic acid is 'washed' from the muscles into the blood.

During recovery lactic acid continues to be 'washed' from muscles but is removed by oxidation in the liver. The additional oxygen needed for this makes up the oxygen debt built up during exercise.

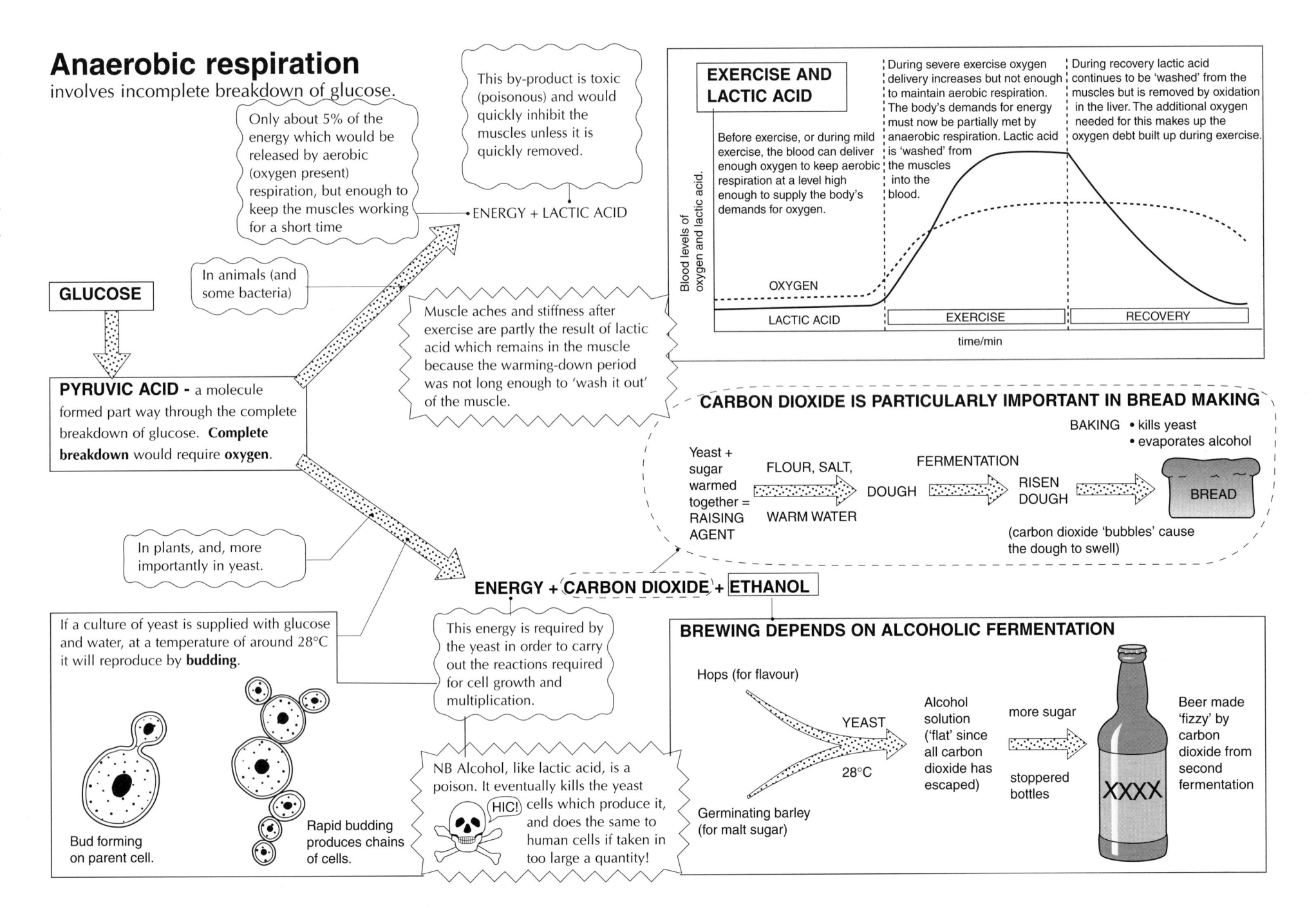

Blood levels of oxygen and lactic acid.

OXYGEN
LACTIC ACID

EXERCISE | RECOVERY
time/min

CARBON DIOXIDE IS PARTICULARLY IMPORTANT IN BREAD MAKING

BAKING • kills yeast
• evaporates alcohol

Yeast + sugar warmed together = RAISING AGENT

FLOUR, SALT, WARM WATER → DOUGH → FERMENTATION → RISEN DOUGH → BREAD

(carbon dioxide 'bubbles' cause the dough to swell)

BREWING DEPENDS ON ALCOHOLIC FERMENTATION

Hops (for flavour)
Germinating barley (for malt sugar)
YEAST 28°C → Alcohol solution ('flat' since all carbon dioxide has escaped) → more sugar, stoppered bottles → Beer made 'fizzy' by carbon dioxide from second fermentation

---

GLUCOSE

This by-product is toxic (poisonous) and would quickly inhibit the muscles unless it is quickly removed.

→ **ENERGY + LACTIC ACID**

Only about 5% of the energy which would be released by aerobic (oxygen present) respiration, but enough to keep the muscles working for a short time

In animals (and some bacteria)

Muscle aches and stiffness after exercise are partly the result of lactic acid which remains in the muscle because the warming-down period was not long enough to 'wash it out' of the muscle.

PYRUVIC ACID - a molecule formed part way through the complete breakdown of glucose. **Complete breakdown** would require **oxygen**.

In plants, and, more importantly in yeast.

→ **ENERGY + CARBON DIOXIDE + ETHANOL**

This energy is required by the yeast in order to carry out the reactions required for cell growth and multiplication.

NB Alcohol, like lactic acid, is a poison. It eventually kills the yeast cells which produce it, and does the same to human cells if taken in too large a quantity!

If a culture of yeast is supplied with glucose and water, at a temperature of around 28°C it will reproduce by **budding**.

Bud forming on parent cell.

Rapid budding produces chains of cells.

54 Anaerobic respiration

Ecology is the study of living organisms in relation to their environment.

→ **THIS MAY INVOLVE**

ADAPTATIONS OF SINGLE ORGANISMS

- Tail fin and muscular tail provide propulsion.
- Spines offer some protection against predators.
- Fins provide stability in water.
- Gills enable uptake of oxygen from water.

These structural adaptations help the individual organism to survive in its **habitat**.

HOW GROUPS OF ORGANISMS INTERACT TO FORM 'UNITS OF THE ENVIRONMENT'

Population - all the members of the same species within a given physical area e.g. all of the 3-spined sticklebacks in a pond.

Community (the **biotic community**) - all of the populations i.e. all of the living organisms within a given physical area e.g. all the animals, plants and micro-organisms within the pond.

Ecosystem - the biotic (living) community together with the physical (non-living or abiotic) environment e.g. a freshwater pond.

OTHER POPULATIONS

PHYSICAL ENVIRONMENT

A **habitat** must provide
- food
- shelter
- breeding sites

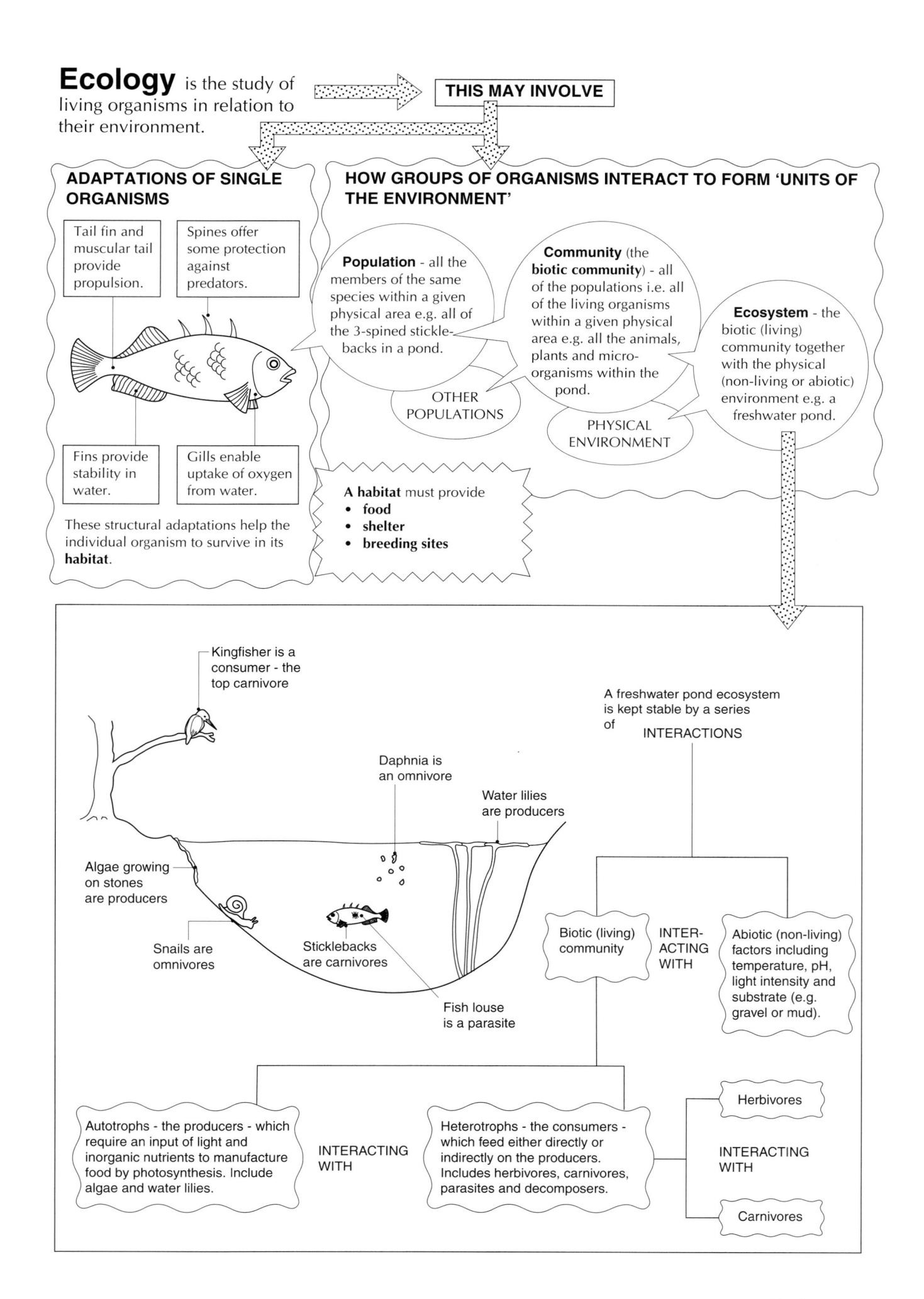

- Kingfisher is a consumer - the top carnivore
- Daphnia is an omnivore
- Water lilies are producers
- Algae growing on stones are producers
- Snails are omnivores
- Sticklebacks are carnivores
- Fish louse is a parasite

A freshwater pond ecosystem is kept stable by a series of **INTERACTIONS**

Biotic (living) community **INTERACTING WITH** Abiotic (non-living) factors including temperature, pH, light intensity and substrate (e.g. gravel or mud).

Autotrophs - the producers - which require an input of light and inorganic nutrients to manufacture food by photosynthesis. Include algae and water lilies.

INTERACTING WITH

Heterotrophs - the consumers - which feed either directly or indirectly on the producers. Includes herbivores, carnivores, parasites and decomposers.

INTERACTING WITH

- Herbivores
- Carnivores

Factors affecting population growth

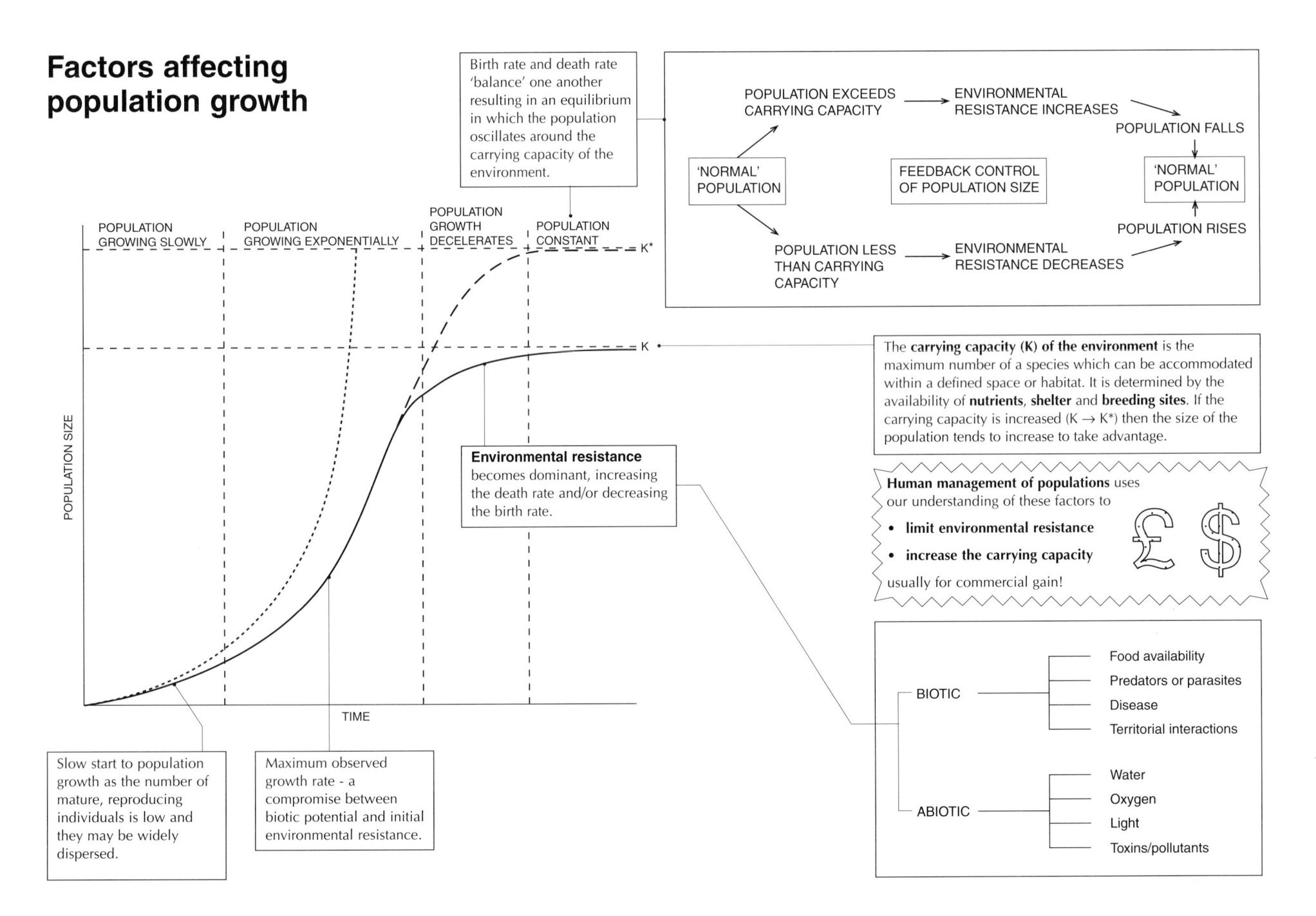

The carrying capacity (K) of the environment is the maximum number of a species which can be accommodated within a defined space or habitat. It is determined by the availability of **nutrients**, **shelter** and **breeding sites**. If the carrying capacity is increased (K → K*) then the size of the population tends to increase to take advantage.

Human management of populations uses our understanding of these factors to
- **limit environmental resistance**
- **increase the carrying capacity**

usually for commercial gain!

BIOTIC
- Food availability
- Predators or parasites
- Disease
- Territorial interactions

ABIOTIC
- Water
- Oxygen
- Light
- Toxins/pollutants

'NORMAL' POPULATION
- POPULATION EXCEEDS CARRYING CAPACITY → ENVIRONMENTAL RESISTANCE INCREASES → POPULATION FALLS
- POPULATION LESS THAN CARRYING CAPACITY → ENVIRONMENTAL RESISTANCE DECREASES → POPULATION RISES

FEEDBACK CONTROL OF POPULATION SIZE → 'NORMAL' POPULATION

Birth rate and death rate 'balance' one another resulting in an equilibrium in which the population oscillates around the carrying capacity of the environment.

Environmental resistance becomes dominant, increasing the death rate and/or decreasing the birth rate.

Maximum observed growth rate - a compromise between biotic potential and initial environmental resistance.

Slow start to population growth as the number of mature, reproducing individuals is low and they may be widely dispersed.

POPULATION GROWING SLOWLY | POPULATION GROWING EXPONENTIALLY | POPULATION GROWTH DECELERATES | POPULATION CONSTANT

K*
K

POPULATION SIZE
TIME

Human population growth

HUMAN SUCCESS
measured as
1. worldwide distribution
2. large number of individuals
3. dominance over other species

largely due to behavioural skills which
a. allow solution of complex problems
b. allows control of/modification of environment

leading to **changes in carrying capacity of the environment**

THREE SIGNIFICANT PHASES IN HUMAN POPULATION GROWTH

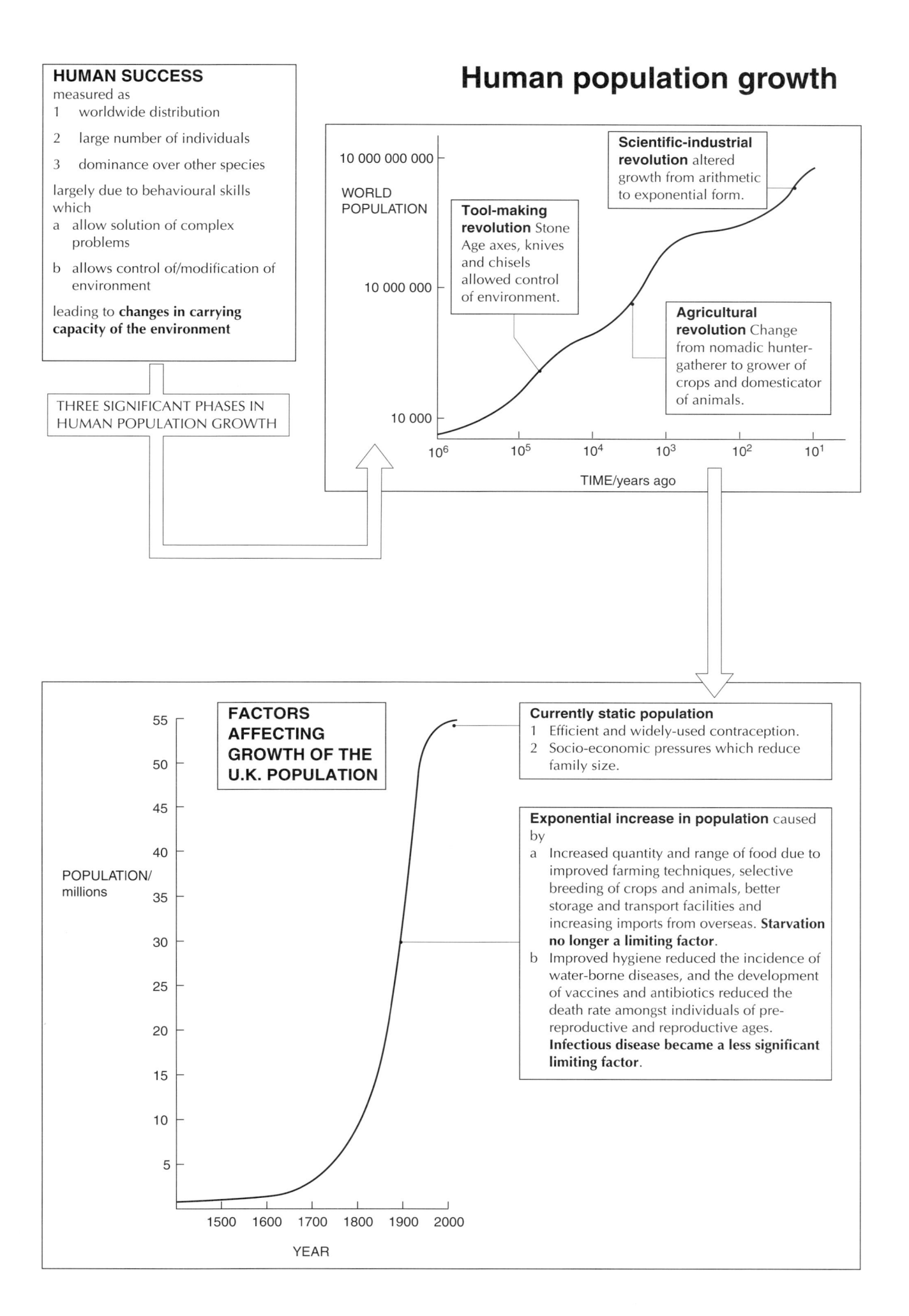

Tool-making revolution Stone Age axes, knives and chisels allowed control of environment.

Agricultural revolution Change from nomadic hunter-gatherer to grower of crops and domesticator of animals.

Scientific-industrial revolution altered growth from arithmetic to exponential form.

(WORLD POPULATION vs TIME/years ago; axes: 10^6, 10^5, 10^4, 10^3, 10^2, 10^1; population: 10 000, 10 000 000, 10 000 000 000)

FACTORS AFFECTING GROWTH OF THE U.K. POPULATION

(POPULATION/millions vs YEAR 1500–2000)

Currently static population
1. Efficient and widely-used contraception.
2. Socio-economic pressures which reduce family size.

Exponential increase in population caused by
a. Increased quantity and range of food due to improved farming techniques, selective breeding of crops and animals, better storage and transport facilities and increasing imports from overseas. **Starvation no longer a limiting factor.**
b. Improved hygiene reduced the incidence of water-borne diseases, and the development of vaccines and antibiotics reduced the death rate amongst individuals of pre-reproductive and reproductive ages. **Infectious disease became a less significant limiting factor.**

Pollution of the atmosphere

LEAD COMPOUNDS may slow mental development. They are in 'anti-knock' additives in petrol, and are released into the atmosphere from exhaust gases. Now most cars use lead-free petrol.

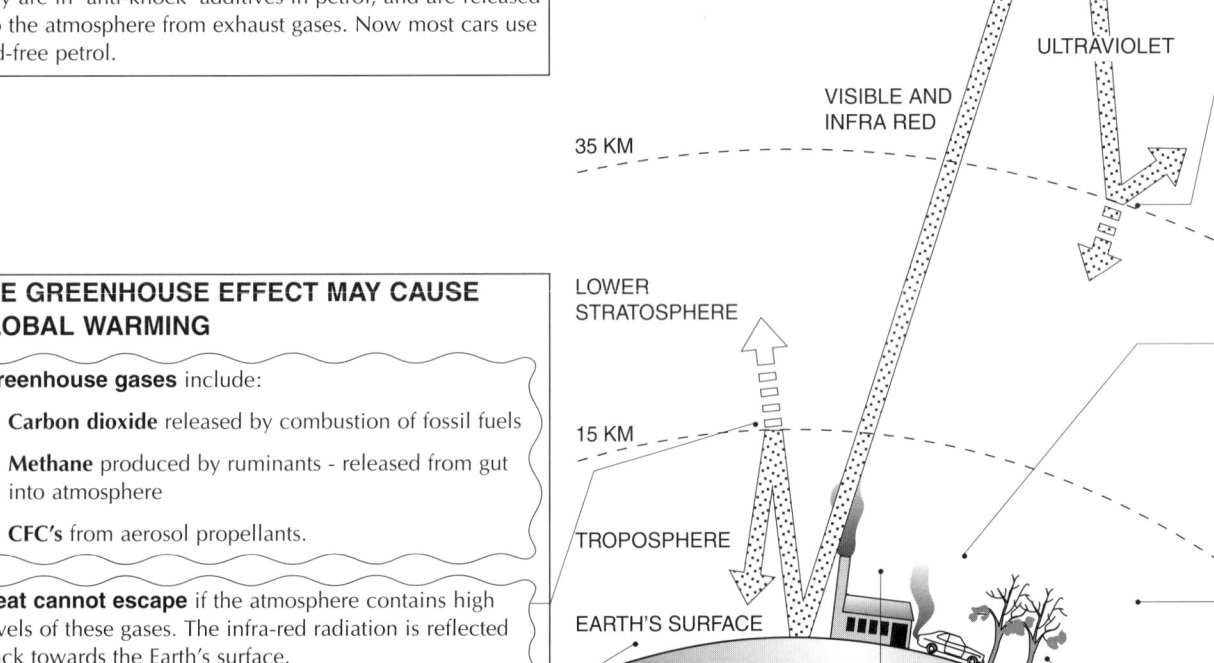

ATMOSPHERIC OZONE IS ESSENTIAL (but in the wrong place it can be harmful!)

High level ozone offers protection: ozone absorbs solar ultraviolet radiation which would reach the Earth's surface and cause
- sunburn
- skin cancer
- cataracts
- mutations (damage to DNA)

'Holes' in the ozone layer have been detected over Antarctica and are thought to have been caused by chlorine from long-lived CFCs.

Human activities affect ozone levels
CFCs (chlorofluorocarbons) used in aerosols, refrigerator coolants and expanded plastics **decrease amounts of high level ozone.**
Fossil fuel combustion produces oxides of nitrogen which react with oxygen **to increase amounts of low level ozone.**

Low level ozone causes problems
- acts as a **greenhouse gas**
- contributes to trapping of dust and smoke: **smog**
- causes **irritation** of eyes, throat and lungs
- damages mesophyll in leaves: **reduction in crops.**

SOLAR RADIATION

ULTRAVIOLET

VISIBLE AND INFRA RED

35 KM

LOWER STRATOSPHERE

15 KM

TROPOSPHERE

EARTH'S SURFACE

THE GREENHOUSE EFFECT MAY CAUSE GLOBAL WARMING

Greenhouse gases include:
- **Carbon dioxide** released by combustion of fossil fuels
- **Methane** produced by ruminants - released from gut into atmosphere
- **CFC's** from aerosol propellants.

Heat cannot escape if the atmosphere contains high levels of these gases. The infra-red radiation is reflected back towards the Earth's surface.

There are good and bad results

✓ more carbon dioxide and higher temperatures mean **more photosynthesis** and **more food.**

✗ Global warming causes
- greater climatic extremes - high winds and heavier rains
- rising sea levels due to melting of polar ice
- crop losses as water evaporates from fertile areas
- extended range of pests

ACID RAIN

Human activities release acidic gases
- Sulphur and nitrogen in fossil fuels are converted to oxides during combustion.
- More oxidation occurs in the clouds, and is catalysed by ozone and unburnt hydrocarbon fuels.
- The oxides dissolve in water, and fall as **acid rain.**

sulphur dioxide and nitrogen oxides $\xrightarrow{H_2O}$ sulphuric and nitric acid

Acid rain causes problems
- **Soils** become very acidic. This causes **leaching of minerals** and **inhibition of decay.**
- **Water** in lakes and rivers collects excess minerals. This causes **death of fish and invertebrates** so that food chains are disrupted.
- **Forest trees** suffer **starvation** because of (a) leaching of ions (b) destruction of photosynthetic tissue.

Acid rain can be reduced
- reduce emissions from car exhausts with **catalytic converters.**
- reduce emissions from power stations with **scrubbers.**
- use 'cleaner' power sources e.g. hydro- and nuclear power.

Pollution of water

PESTICIDES – overuse of pesticides on agricultural land (e.g. to protect a crop from insects) or directly on water (e.g. to kill an aquatic stage of an insect) can raise pesticide levels. The pesticide levels are then **amplified** as they pass through food chains

e.g. DDT concentration in parts per million:

WATER 0.02 → PLANKTON 5 → DAPHNIA 50 → STICKLEBACK 250 → GREBE 1500

At these concentrations the DDT is harmful and reduces breeding success.

THERMAL POLLUTION – industries / power stations use water as a coolant, then discharge the water into rivers.

Now has **lower oxygen concentration** as the capacity of water for dissolved oxygen decreases as temperature rises, and as fish and bacteria become more active and so respire more. May have **new species** e.g. tropical fish, terrapins which may affect food chains.

LEAD POLLUTION – water pipes were traditionally made of lead. Lead dissolves into the water which flows through the pipes. Lead compounds are toxic and accumulate via food chains. Lead weights discarded by anglers also contribute to lead pollution.

OIL POLLUTION – oil tankers spill their contents, by accident or deliberately (during cleaning), into the sea. The oil floats on the surface, causing
- death of seabirds since feathers lose their ability to insulate when they are coated with oil.
- fish are directly poisoned
- marine mammals are killed by eating poisoned food or by loss of fur's insulating capacity.

B.O.D. is the mass of oxygen consumed by micro-organisms in a sample of water. It is determined by measuring oxygen concentration with an oxygen electrode **before** and **after** a period of microbial respiration. It indicates the oxygen **not available** to more advanced organisms.

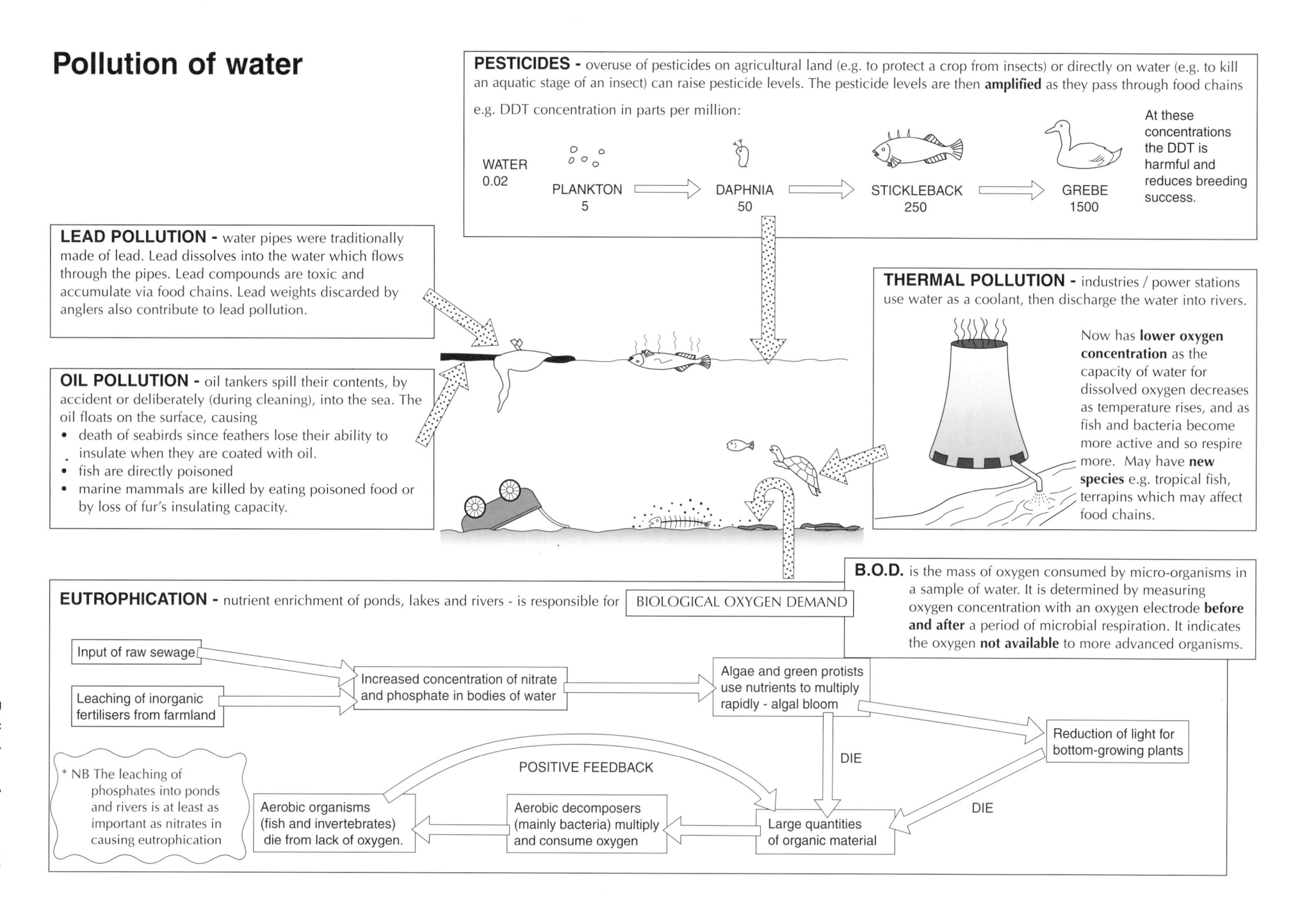

BIOLOGICAL OXYGEN DEMAND

EUTROPHICATION – nutrient enrichment of ponds, lakes and rivers – is responsible for

Input of raw sewage → Increased concentration of nitrate and phosphate in bodies of water

Leaching of inorganic fertilisers from farmland → Increased concentration of nitrate and phosphate in bodies of water

Increased concentration of nitrate and phosphate in bodies of water → Algae and green protists use nutrients to multiply rapidly – algal bloom

Algae and green protists... → Reduction of light for bottom-growing plants → DIE → Large quantities of organic material

Algae... DIE → Large quantities of organic material

Large quantities of organic material → Aerobic decomposers (mainly bacteria) multiply and consume oxygen → Aerobic organisms (fish and invertebrates) die from lack of oxygen.

Aerobic organisms die → Large quantities of organic material (POSITIVE FEEDBACK)

* NB The leaching of phosphates into ponds and rivers is at least as important as nitrates in causing eutrophication

Saprotrophs cause decay

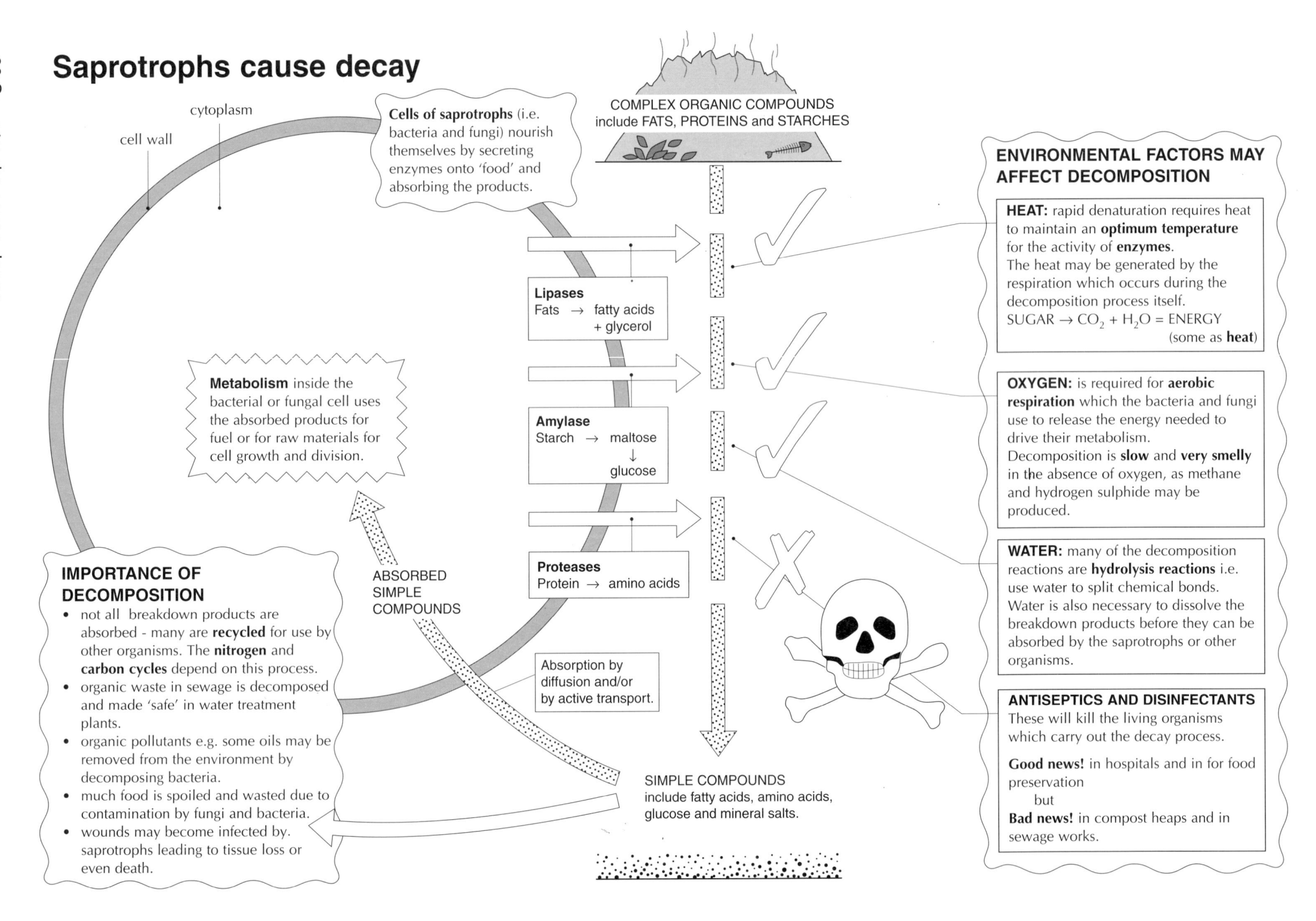

Cells of saprotrophs (i.e. bacteria and fungi) nourish themselves by secreting enzymes onto 'food' and absorbing the products.

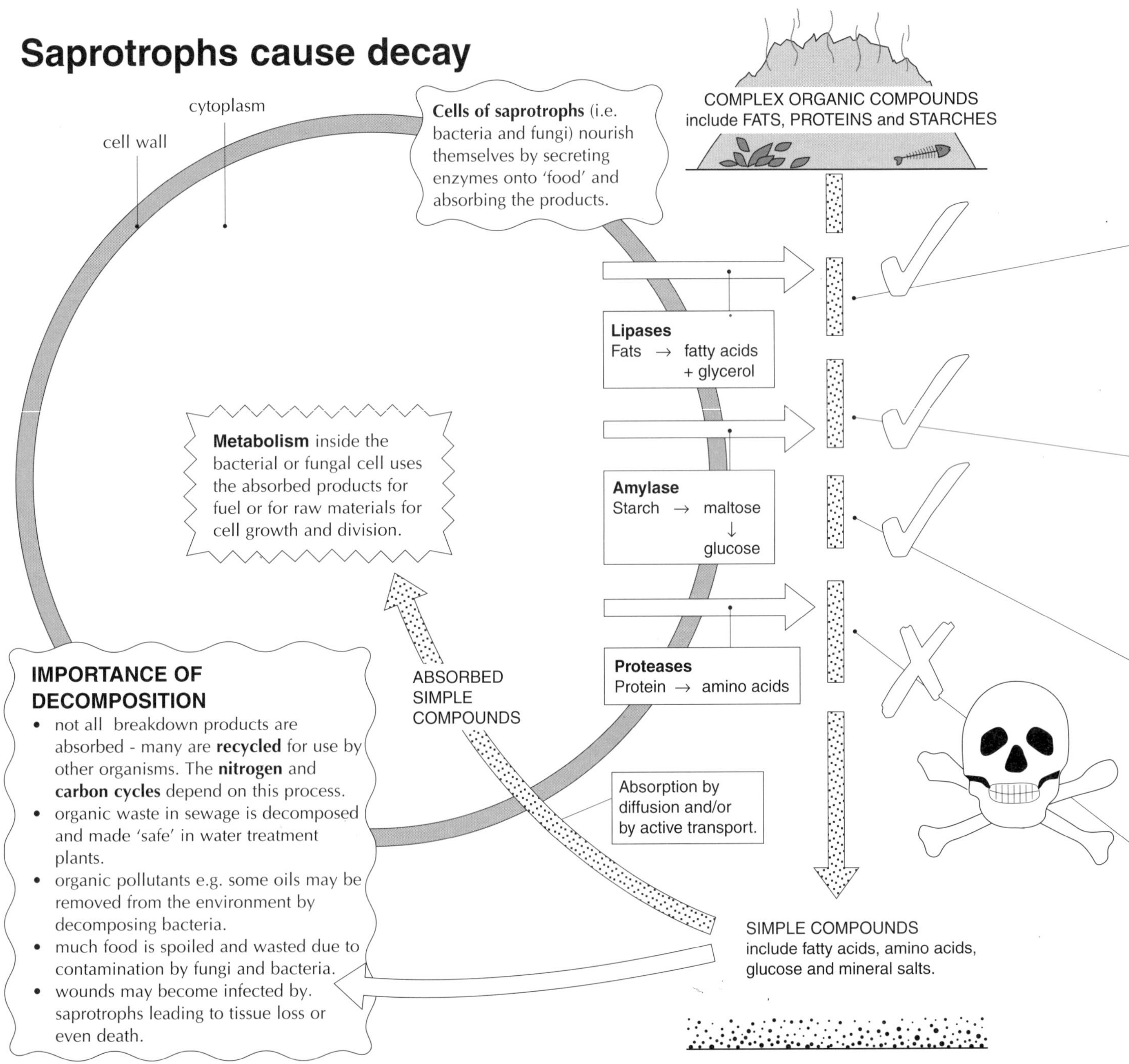

- cytoplasm
- cell wall

COMPLEX ORGANIC COMPOUNDS include FATS, PROTEINS and STARCHES

Lipases Fats → fatty acids + glycerol

Amylase Starch → maltose → glucose

Proteases Protein → amino acids

Absorption by diffusion and/or by active transport.

SIMPLE COMPOUNDS include fatty acids, amino acids, glucose and mineral salts.

ABSORBED SIMPLE COMPOUNDS

Metabolism inside the bacterial or fungal cell uses the absorbed products for fuel or for raw materials for cell growth and division.

ENVIRONMENTAL FACTORS MAY AFFECT DECOMPOSITION

HEAT: rapid denaturation requires heat to maintain an **optimum temperature** for the activity of **enzymes**.
The heat may be generated by the respiration which occurs during the decomposition process itself.
SUGAR → CO_2 + H_2O = ENERGY
(some as **heat**)

OXYGEN: is required for **aerobic respiration** which the bacteria and fungi use to release the energy needed to drive their metabolism.
Decomposition is **slow** and **very smelly** in the absence of oxygen, as methane and hydrogen sulphide may be produced.

WATER: many of the decomposition reactions are **hydrolysis reactions** i.e. use water to split chemical bonds. Water is also necessary to dissolve the breakdown products before they can be absorbed by the saprotrophs or other organisms.

ANTISEPTICS AND DISINFECTANTS
These will kill the living organisms which carry out the decay process.
Good news! in hospitals and in for food preservation
but
Bad news! in compost heaps and in sewage works.

IMPORTANCE OF DECOMPOSITION

- not all breakdown products are absorbed - many are **recycled** for use by other organisms. The **nitrogen** and **carbon cycles** depend on this process.
- organic waste in sewage is decomposed and made 'safe' in water treatment plants.
- organic pollutants e.g. some oils may be removed from the environment by decomposing bacteria.
- much food is spoiled and wasted due to contamination by fungi and bacteria.
- wounds may become infected by saprotrophs leading to tissue loss or even death.

Treatment of sewage

Treatment of sewage uses physical and biological processes to limit damage to the aquatic environment.

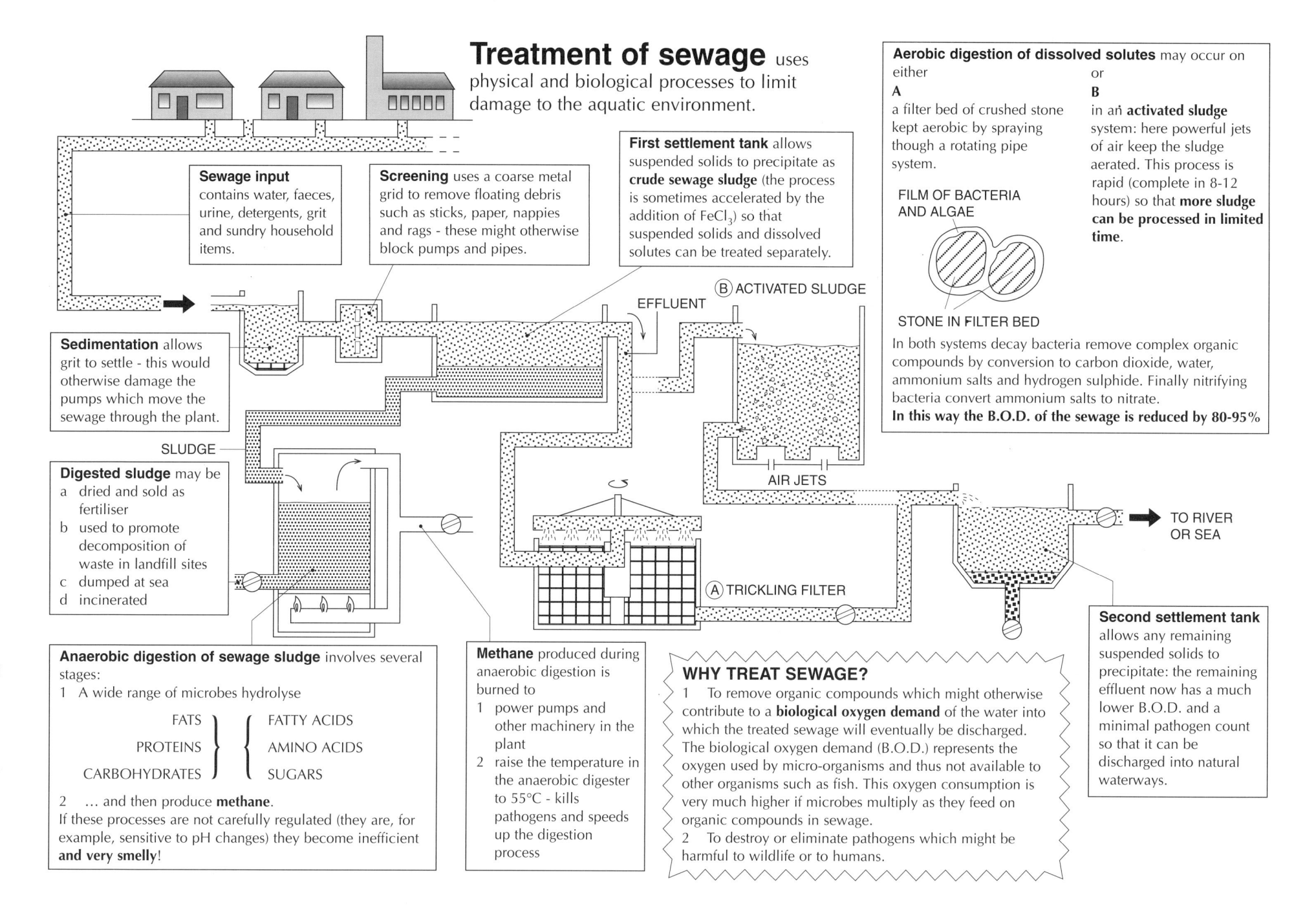

Sewage input contains water, faeces, urine, detergents, grit and sundry household items.

Screening uses a coarse metal grid to remove floating debris such as sticks, paper, nappies and rags - these might otherwise block pumps and pipes.

Sedimentation allows grit to settle - this would otherwise damage the pumps which move the sewage through the plant.

First settlement tank allows suspended solids to precipitate as **crude sewage sludge** (the process is sometimes accelerated by the addition of $FeCl_3$) so that suspended solids and dissolved solutes can be treated separately.

Aerobic digestion of dissolved solutes may occur on either

A
a filter bed of crushed stone kept aerobic by spraying though a rotating pipe system.

FILM OF BACTERIA AND ALGAE
STONE IN FILTER BED

or

B
in an **activated sludge** system: here powerful jets of air keep the sludge aerated. This process is rapid (complete in 8-12 hours) so that **more sludge can be processed in limited time**.

In both systems decay bacteria remove complex organic compounds by conversion to carbon dioxide, water, ammonium salts and hydrogen sulphide. Finally nitrifying bacteria convert ammonium salts to nitrate.

In this way the B.O.D. of the sewage is reduced by 80-95%

(A) TRICKLING FILTER
(B) ACTIVATED SLUDGE
AIR JETS
EFFLUENT
SLUDGE
TO RIVER OR SEA

Second settlement tank allows any remaining suspended solids to precipitate: the remaining effluent now has a much lower B.O.D. and a minimal pathogen count so that it can be discharged into natural waterways.

Methane produced during anaerobic digestion is burned to
1 power pumps and other machinery in the plant
2 raise the temperature in the anaerobic digester to 55°C - kills pathogens and speeds up the digestion process

Anaerobic digestion of sewage sludge involves several stages:
1 A wide range of microbes hydrolyse

FATS → FATTY ACIDS
PROTEINS → AMINO ACIDS
CARBOHYDRATES → SUGARS

2 ... and then produce **methane**.

If these processes are not carefully regulated (they are, for example, sensitive to pH changes) they become inefficient and **very smelly**!

Digested sludge may be
a dried and sold as fertiliser
b used to promote decomposition of waste in landfill sites
c dumped at sea
d incinerated

WHY TREAT SEWAGE?
1 To remove organic compounds which might otherwise contribute to a **biological oxygen demand** of the water into which the treated sewage will eventually be discharged. The biological oxygen demand (B.O.D.) represents the oxygen used by micro-organisms and thus not available to other organisms such as fish. This oxygen consumption is very much higher if microbes multiply as they feed on organic compounds in sewage.
2 To destroy or eliminate pathogens which might be harmful to wildlife or to humans.

Ecological pyramids

represent numerical relationships between successive trophic levels in a community.

PYRAMID OF NUMBERS
a diagrammatic representation of the numbers of different organisms at each trophic level in an ecosystem **at any one time**.

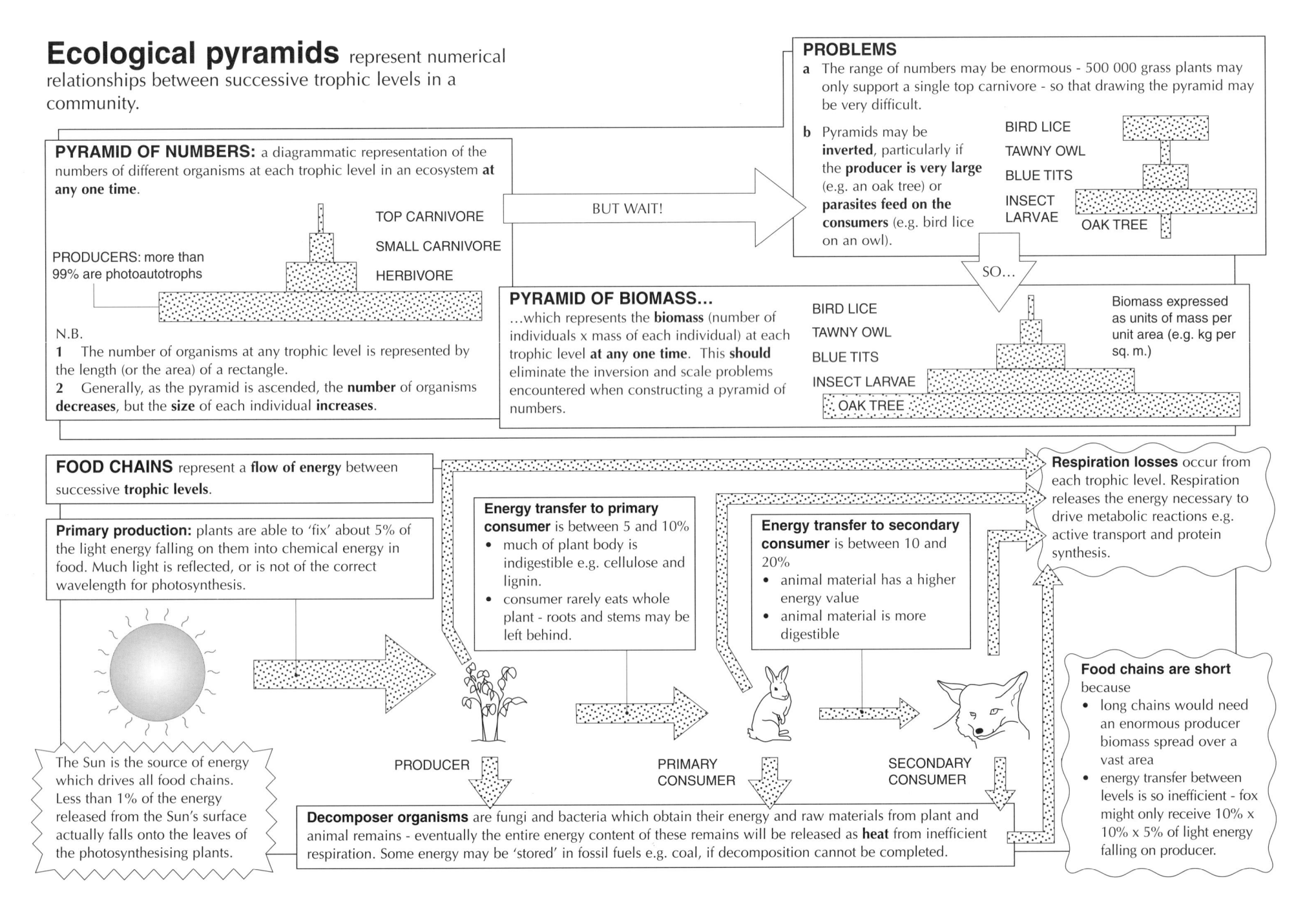

TOP CARNIVORE
SMALL CARNIVORE
HERBIVORE
PRODUCERS: more than 99% are photoautotrophs

N.B.
1. The number of organisms at any trophic level is represented by the length (or the area) of a rectangle.
2. Generally, as the pyramid is ascended, the **number** of organisms **decreases**, but the **size** of each individual **increases**.

PROBLEMS

a. The range of numbers may be enormous – 500 000 grass plants may only support a single top carnivore – so that drawing the pyramid may be very difficult.

b. Pyramids may be **inverted**, particularly if the **producer is very large** (e.g. an oak tree) or **parasites feed on the consumers** (e.g. bird lice on an owl).

BIRD LICE
TAWNY OWL
BLUE TITS
INSECT LARVAE
OAK TREE

BUT WAIT!

SO...

PYRAMID OF BIOMASS...

...which represents the **biomass** (number of individuals x mass of each individual) at each trophic level **at any one time**. This **should** eliminate the inversion and scale problems encountered when constructing a pyramid of numbers.

BIRD LICE
TAWNY OWL
BLUE TITS
INSECT LARVAE
OAK TREE

Biomass expressed as units of mass per unit area (e.g. kg per sq. m.)

FOOD CHAINS
represent a **flow of energy** between successive **trophic levels**.

Primary production: plants are able to 'fix' about 5% of the light energy falling on them into chemical energy in food. Much light is reflected, or is not of the correct wavelength for photosynthesis.

The Sun is the source of energy which drives all food chains. Less than 1% of the energy released from the Sun's surface actually falls onto the leaves of the photosynthesising plants.

PRODUCER → PRIMARY CONSUMER → SECONDARY CONSUMER

Energy transfer to primary consumer is between 5 and 10%
- much of plant body is indigestible e.g. cellulose and lignin.
- consumer rarely eats whole plant – roots and stems may be left behind.

Energy transfer to secondary consumer is between 10 and 20%
- animal material has a higher energy value
- animal material is more digestible

Respiration losses occur from each trophic level. Respiration releases the energy necessary to drive metabolic reactions e.g. active transport and protein synthesis.

Food chains are short because
- long chains would need an enormous producer biomass spread over a vast area
- energy transfer between levels is so inefficient – fox might only receive 10% x 10% x 5% of light energy falling on producer.

Decomposer organisms are fungi and bacteria which obtain their energy and raw materials from plant and animal remains – eventually the entire energy content of these remains will be released as **heat** from inefficient respiration. Some energy may be 'stored' in fossil fuels e.g. coal, if decomposition cannot be completed.

62 Ecological pyramids

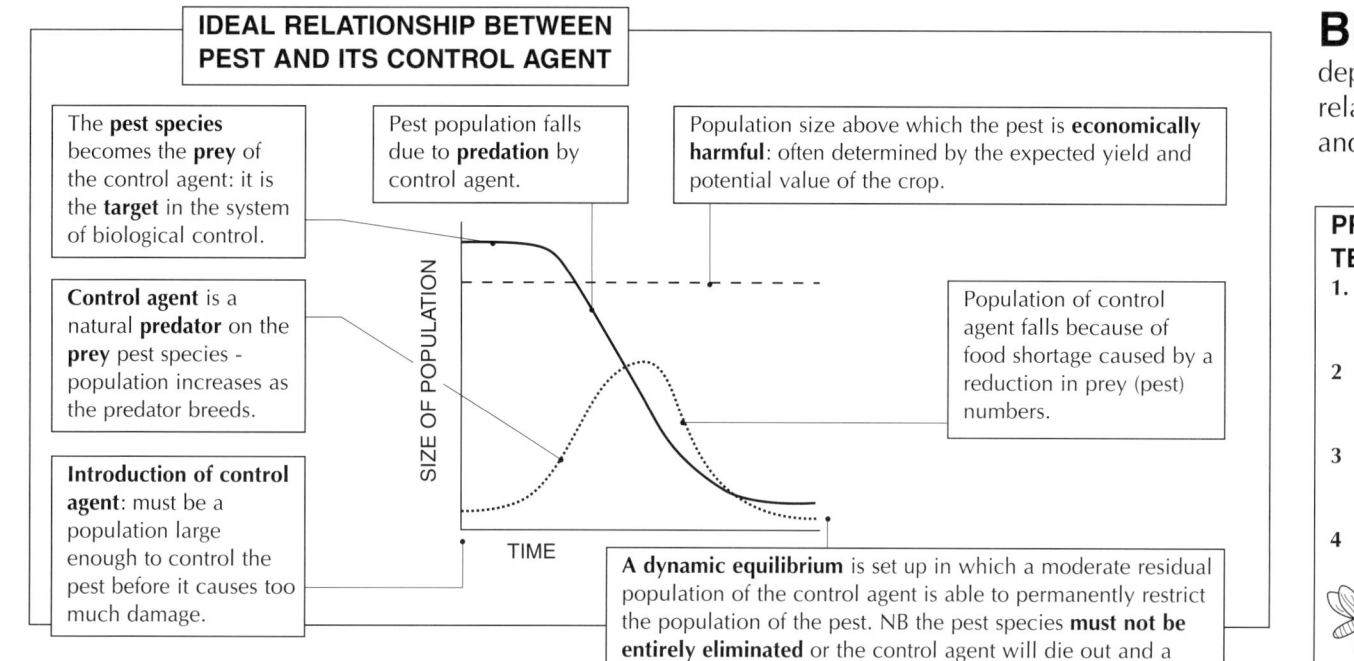

Biological pest control

depends on an understanding of the relationship between populations of predators and their prey.

PRINCIPLE
TECHNIQUES IN BIOLOGICAL CONTROL

1. Use a **herbivore** to control a **weed** species e.g. *Cactoblastis* larvae on prickly pear.
2. Use a **carnivore** to control an **herbivorous pest** e.g. hoverfly larvae on aphids.
3. Use a **parasite** to control its **host** e.g. *Encharsia*, a parasitic wasp, on the greenhouse whitefly, *Trialeurodes vaporariorum*.
4. Disrupt the **breeding cycle** of a pest **if it mates once only in its life** e.g. release of irradiation-sterilised males of the screw worm fly, a flesh eating parasite of cattle.
5. Control of **pest behaviour** e.g. sex attractant pheromones are used to attract apple codling moths into lethal traps.

IDEAL RELATIONSHIP BETWEEN PEST AND ITS CONTROL AGENT

The **pest species** becomes the **prey** of the control agent: it is the **target** in the system of biological control.

Control agent is a natural **predator** on the prey species - population increases as the predator breeds.

Introduction of control agent: must be a population large enough to control the pest before it causes too much damage.

Pest population falls due to **predation** by control agent.

Population size above which the pest is **economically harmful**: often determined by the expected yield and potential value of the crop.

Population of control agent falls because of food shortage caused by a reduction in prey (pest) numbers.

A dynamic equilibrium is set up in which a moderate residual population of the control agent is able to permanently restrict the population of the pest. NB the pest species **must not be entirely eliminated** or the control agent will die out and a further introduction will be necessary to prevent re-establishment of economically damaging pest populations.

SIZE OF POPULATION / TIME

... BUT THERE HAVE BEEN FAILURES!

Hawaiian cane toads introduced into Queensland, Australia, to control the greyback beetle, a pest on sugar cane, are now a serious threat to Australian wildlife. The toad
- eats many native insects and worms
- displaces native frogs and toads from breeding pools
- poisons larger animals which try to eat it because its skin is extremely toxic.

Cats and stoats introduced to offshore islands of Britain and New Zealand to limit population of rodents which threatened rare ground-nesting birds found that the chicks of the 'protected' birds were easier to catch than the rats were!

NEVERTHELESS, COMPARED WITH CHEMICAL CONTROL, THERE ARE CERTAIN ADVANTAGES OF BIOLOGICAL PEST CONTROL...

	BIOLOGICAL CONTROL	CHEMICAL CONTROL
SPECIFICITY	Very high, so little danger to beneficial species or to humans	May be low, but legislation leading to great improvements
'ACCUMULATION' IN ECOSYSTEMS	None	Concentrations may increase along food chains
PERMANENCE OF CONTROL	Good, but small numbers of pests must be tolerated	Requires regular reapplication
DEVELOPMENT OF PEST RESISTANCE	Very rare	Common, requiring ever-increasing 'dose'
COST IN FINANCIAL TERMS	Initially may be high, but very low in the long-term	May be very high, restricting use to wealthy nations

Cycling of nutrients involves interconversion of simple and complex molecules

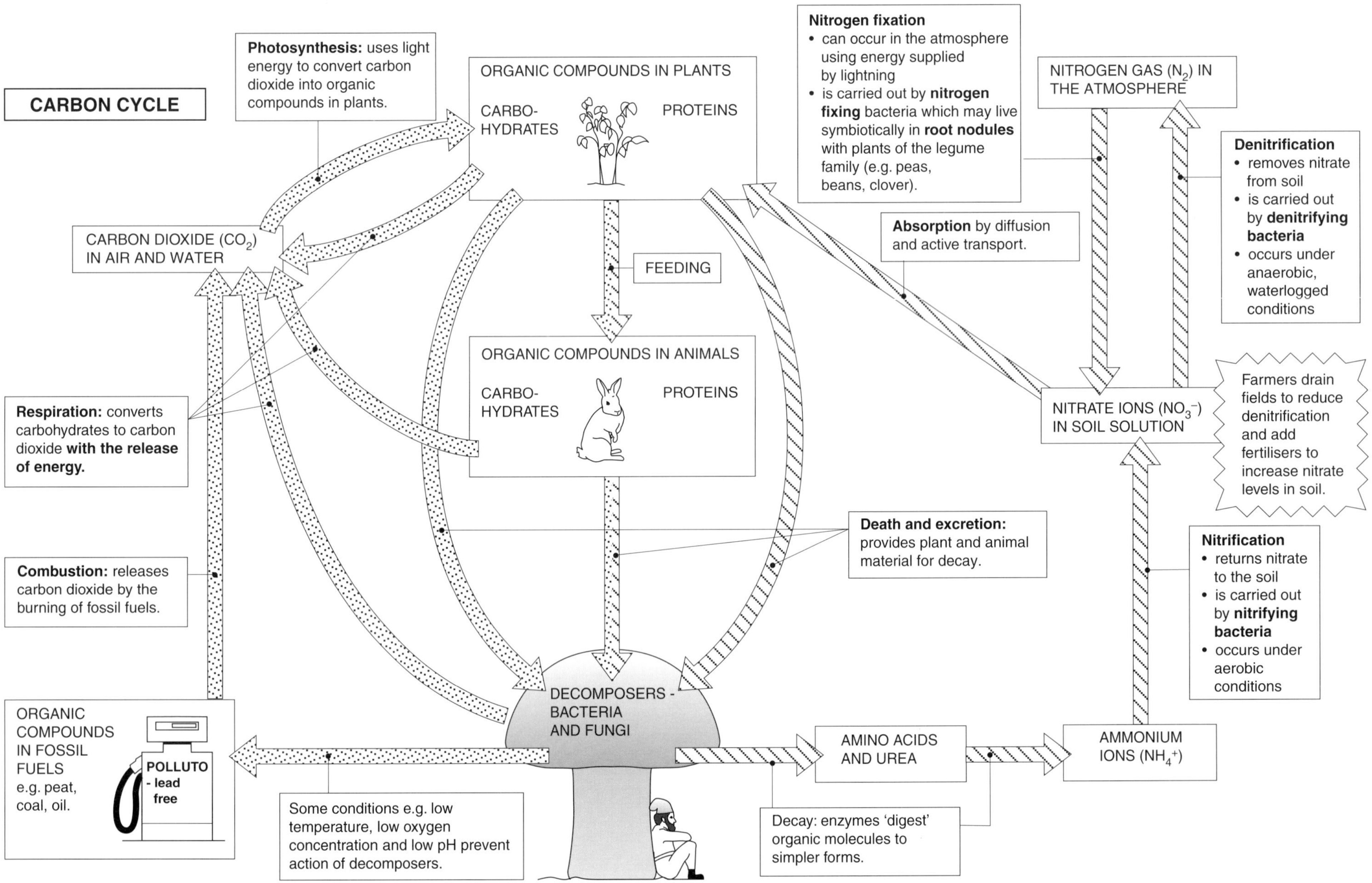

NITROGEN CYCLE

Denitrification
- removes nitrate from soil
- is carried out by **denitrifying bacteria**
- occurs under anaerobic, waterlogged conditions

Nitrification
- returns nitrate to the soil
- is carried out by **nitrifying bacteria**
- occurs under aerobic conditions

Farmers drain fields to reduce denitrification and add fertilisers to increase nitrate levels in soil.

NITROGEN GAS (N_2) IN THE ATMOSPHERE

NITRATE IONS (NO_3^-) IN SOIL SOLUTION

AMMONIUM IONS (NH_4^+)

Nitrogen fixation
- can occur in the atmosphere using energy supplied by lightning
- is carried out by **nitrogen fixing** bacteria which may live symbiotically in **root nodules** with plants of the legume family (e.g. peas, beans, clover).

Absorption by diffusion and active transport.

Death and excretion: provides plant and animal material for decay.

AMINO ACIDS AND UREA

Decay: enzymes 'digest' organic molecules to simpler forms.

ORGANIC COMPOUNDS IN PLANTS
- PROTEINS
- CARBOHYDRATES

FEEDING

ORGANIC COMPOUNDS IN ANIMALS
- PROTEINS
- CARBOHYDRATES

DECOMPOSERS - BACTERIA AND FUNGI

Some conditions e.g. low temperature, low oxygen concentration and low pH prevent action of decomposers.

CARBON CYCLE

Photosynthesis: uses light energy to convert carbon dioxide into organic compounds in plants.

CARBON DIOXIDE (CO_2) IN AIR AND WATER

Respiration: converts carbohydrates to carbon dioxide **with the release of energy.**

Combustion: releases carbon dioxide by the burning of fossil fuels.

POLLUTO - lead free

ORGANIC COMPOUNDS IN FOSSIL FUELS e.g. peat, coal, oil.

Managing ecosystems: fish farming

maximises profit by minimising environmental resistance

PREPARED FOOD

- dried, in pellet form, for convenience of transport from site of production and for measured delivery to fish.
- usually made from 'trash fish' which is marine fish caught in nets but not used for human consumption – **very high in protein** (rapid growth rate) but **very expensive**.
- may include **colouring agent** (turns fish pink – favoured by consumers) and **antibiotic** (disease control).

HANGING NET

- keeps out **aquatic predators** such as otter and pike
- keeps out other fish and so **reduces competition** for the pelleted food.

SUSPENDED NET

- keeps out **aerial predators** such as cormorant, heron and osprey.

PROBLEMS AND ENVIRONMENTAL CONCERNS

- very high food costs.
- poor control of temperature and oxygen availability in large outdoor farming pens.
- much more research necessary to obtain highest yields – particularly important are **selective breeding** programmes to develop new fish varieties with improved growth rates and conversion ratios.

△ pollution by pesticides since these compounds may kill organisms which are foods for 'wild' species.

△ excess food and fish faeces create nutrient rich environment below netted area → growth of bacterial population → increased **biological oxygen demand**.

In developed countries intensive fish farming is so costly that it should be seen as **adding variety to the diet** rather than being a **source of cheap, plentiful protein**.

DISEASE CONTROL
close confinement makes fish more likely to suffer from disease ('spread' is much easier)
- dose water with **dichlorvos** which kills **fish lice**.
- dose water with **fungicide** to prevent **fungal** infection of skin and gills.
- add **antibiotic** to food to control **bacterial** infections.

AGE FOR HARVESTING

Harvest at this point: fish still have high conversion ratio so cost of production has been kept to a minimum.

MASS / AGE

SPECIES – the 'farmed' species must

- grow well under captive conditions
- accept prepared i.e. non-living food
- have a high **conversion ratio** i.e. convert food → flesh efficiently
- ideally be able to complete its life cycle i.e. breed under captive conditions
- not be particularly susceptible to disease.

It is usually **expensive** to provide these conditions so the product should **command a high price**. For this reason the main farmed species **in Britain** are members of the Salmon family.

FRY PRODUCTION

- fish are spawned under artificial conditions - often in aquaria and often under the influence of reproductive hormones.
- fry are 'grown on' to a size at which they can be released
  - control temperature (higher temperature, rapid growth)
  - high oxygen levels with aerators
  - growth hormones may be added to water
- fry of **uniform size** are released into the farming pens. This reduces the chance of the fish eating each other!

Managing ecosystems: horticulture

can be profitable if environmental resistance is reduced

CARBON DIOXIDE CONCENTRATION: is a major limiting factor in photosynthesis. In a greenhouse the plants may photosynthesise very quickly and CO_2 is rapidly used up. The CO_2 concentration is usually raised to about 0.1% of the atmosphere (about three times that in air). This gives a significant (about 50%) increase in crop yield. The extra CO_2 can be provided by burning paraffin (which also raises the temperature) or, more accurately, by releasing it from a cylinder.

HUMIDITY: this affects the rate of transpiration, and therefore the rate at which the plants must be supplied with water. Thus a high humidity reduces the need for additional water and generally favours growth (NB too high humidity favours the growth of fungal pests). Humidity is reduced by opening ventilators and raised using an automatic mist spray.

ILLUMINATION: it is important to control
- **intensity** - this will influence the rate of food production by photosynthesis. Greater intensity → more photosynthesis until some other limiting factor intervenes.
- **quality** - photosynthesis is most efficient at red and blue wavelengths so that 'white' light contains some wavelengths ('green') which are not useful.
- **duration** - if fruit is the desired product the plant must flower. Flowering is controlled by daylength (the duration of light in a 24 hour period). Sunlight provides some illumination but artificial lighting systems are more controllable (but more expensive).

PEST CONTROL: is particularly important since a pest species could spread very rapidly through a greenhouse containing only a single crop.
- **Fungi** can be a problem since high humidity/high temperature encourages spore germination and growth of fungus. Control with **fungicide spray** or by **reduction of humidity**.
- **Weed plants** could compete with the crop species for light, water and mineral nutrients. They are removed **by hand** or with a **selective herbicide**.
- **Animal pests** are herbivorous (e.g. caterpillars) or sap-sucking (e.g. aphids) insects. These may be controlled using **insecticides** but
  - these can be inefficient, especially against waxy-coated pupae
  - they may need to be reapplied, so can be expensive
  - they may leave residues so that crop will need washing

  or **biological control** which
  - uses a natural predator of the pest
  - usually only requires a single application
  - does not leave residues

  Important examples of biological control are **ladybirds** which eat aphids **wasp larvae** which live as parasites on **whitefly larvae**

POLLINATING INSECTS: bees may be introduced to ensure high rates of fertilisation and therefore fruit formation.

COMPUTER CONTROL: this is widely used in large commercial greenhouses. Sensors provide information about air temperature, CO_2 concentration, water available to roots, humidity and mineral concentration, and the control centre ensures that any changes are corrected.

e.g. temperature → sensor → heater

HIGH-YIELDING STRAINS OF CROP: selective breeding and/or genetic engineering can develop crop strains which
- have a high yield
- produce fruit of a desirable colour/texture
- produce fruit at the same time
- may have genetic pest resistance.

TEMPERATURE: affects plant growth because of its affect on the enzymes of photosynthesis. NB high temperatures may also speed up the life cycle of pests. Sunlight provides some heat (shading may be necessary in summer) but more control is available using thermostatically-regulated heaters.

Managing ecosystems: animal husbandry.

Intensive farming aims to minimise cost/maximise productivity by reducing environmental resistance.

GROWTH CURVE may indicate optimum time for marketing animals

Sell at this point as 'conversion' begins to fall off: about 100 days for veal

MASS / TIME

Temperature control: essential so that costly heat energy is not wasted.
if **too high**, animals are uncomfortable and will not feed
if **too low**, food intake is 'wasted' on heat production to maintain body temperature.

Controlled light regime/photoperiod: may influence growth rate since can permit longer feeding period
may control reproductive cycle so, for example, milk production may be stabilised through the year.

Shelter:
Prevent entry of **predators**
Eliminate/control **competitors** for food
Protect against **climatic extremes**

Food input:
Control content - high protein for growth
- minimal fat to suit customer demand for lean meat
- include growth hormone to increase growth rate
- add copper ions which reduce energy consumption for heat production

Often use dried milk powder or single cell protein, with mineral and vitamin supplements.

Strain of animal: selective breeding for animals with high **conversion ratio** i.e. most efficient transfer of food intake to body mass.

Minimise movement: less energy consumption and thus more efficient 'conversion'.
NB can upset **social interactions**, and stress can lead to poor growth/unpleasant 'cortisone' taste.

Veterinary care:
Antibiotics to reduce bacterial infections
Vaccination to minimise viral infections
Hormone/vitamin supplements can be administered more accurately than in the diet
Artificial insemination techniques can reduce costs (no need to keep bulls in dairy farms).

Hygienic conditions: most animals are free of gut parasites and debilitating bacteria - healthy animals grow more quickly and meat is more saleable.
Slurry (faeces/urine) can be dried and recycled for use as fertiliser.

Managing ecosystems: animal husbandry 67

Design an experiment

An experiment is designed to **test the validity of an hypothesis** and involves the **collection of data**

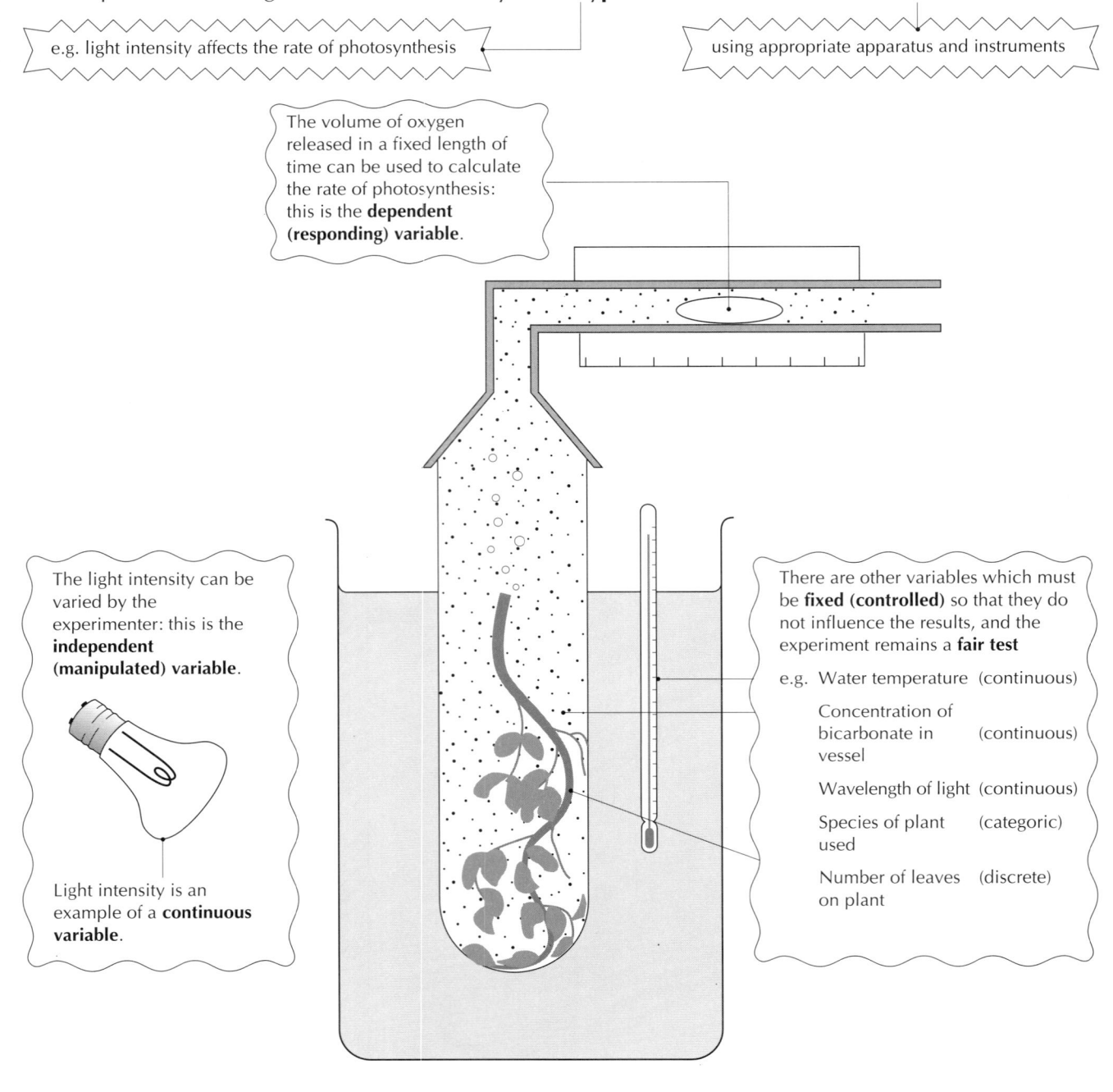

e.g. light intensity affects the rate of photosynthesis

using appropriate apparatus and instruments

The volume of oxygen released in a fixed length of time can be used to calculate the rate of photosynthesis: this is the **dependent (responding) variable**.

The light intensity can be varied by the experimenter: this is the **independent (manipulated) variable**.

Light intensity is an example of a **continuous variable**.

There are other variables which must be **fixed (controlled)** so that they do not influence the results, and the experiment remains a **fair test**

e.g. Water temperature (continuous)

Concentration of bicarbonate in vessel (continuous)

Wavelength of light (continuous)

Species of plant used (categoric)

Number of leaves on plant (discrete)

A **control** experiment is the same in every respect **except** the manipulated variable is not changed but is kept constant.

A control allows confirmation that **no unknown variable** is responsible for any observed changes in the responding variable: it helps to make the experiment a **fair test**.

A '**repeat**' is performed when the experimenter suspects that misleading data has been obtained through "operator error".

'**Means**' of a series of results minimise the influence of any single result, and therefore reduce the effect of any 'rogue' or anomalous data.

68 Design an experiment

Dealing with data

may involve a number of steps

1. Organisation of the **raw data** (the information which you actually collect during your investigation).
2. Manipulation of the data (converting your measurements into another form).

and

3. Representation of the data in **graphical** or other form.

Steps 1 and 2 usually involve **preparation of a table of results**

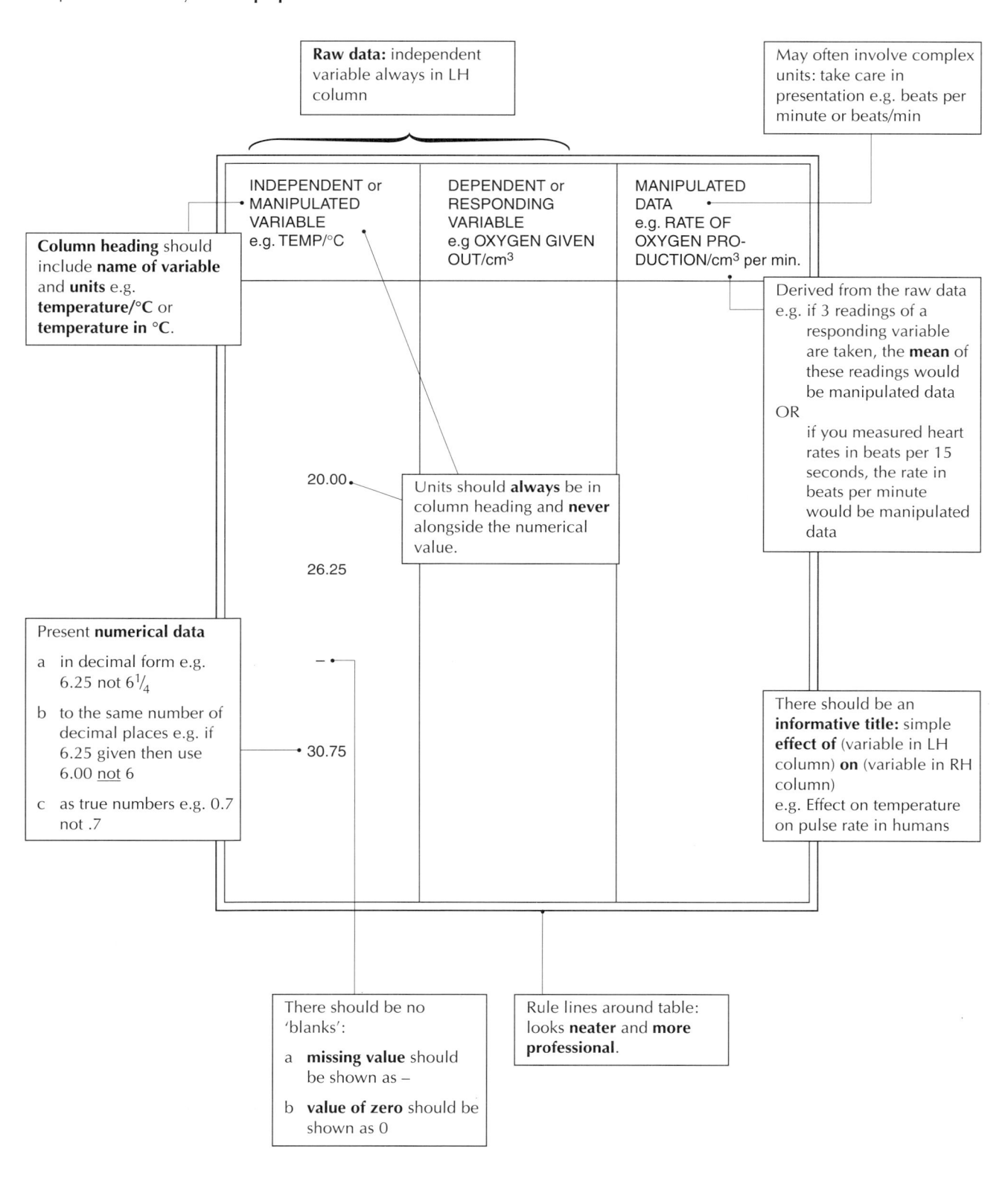

Raw data: independent variable always in LH column

May often involve complex units: take care in presentation e.g. beats per minute or beats/min

Column heading should include **name of variable** and **units** e.g. **temperature/°C** or **temperature in °C**.

INDEPENDENT or MANIPULATED VARIABLE e.g. TEMP/°C	DEPENDENT or RESPONDING VARIABLE e.g OXYGEN GIVEN OUT/cm³	MANIPULATED DATA e.g. RATE OF OXYGEN PRODUCTION/cm³ per min.
20.00		
26.25		
–		
30.75		

Units should **always** be in column heading and **never** alongside the numerical value.

Derived from the raw data e.g. if 3 readings of a responding variable are taken, the **mean** of these readings would be manipulated data

OR

if you measured heart rates in beats per 15 seconds, the rate in beats per minute would be manipulated data

Present **numerical data**

a. in decimal form e.g. 6.25 not 6¼

b. to the same number of decimal places e.g. if 6.25 given then use 6.00 not 6

c. as true numbers e.g. 0.7 not .7

There should be an **informative title**: simple **effect of** (variable in LH column) **on** (variable in RH column) e.g. Effect on temperature on pulse rate in humans

There should be no 'blanks':

a. **missing value** should be shown as –

b. **value of zero** should be shown as 0

Rule lines around table: looks **neater** and **more professional**.

Graphical representation:
A graph is a visual presentation of data and may help to make the relationship between variables more obvious.

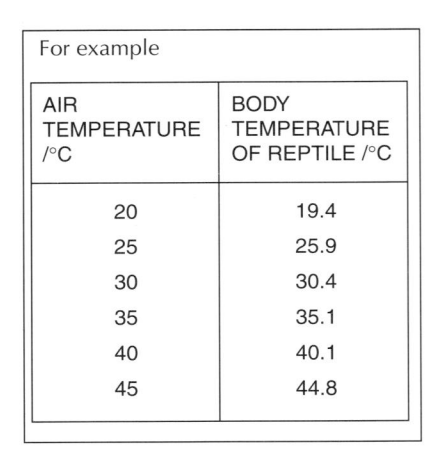

For example

AIR TEMPERATURE /°C	BODY TEMPERATURE OF REPTILE /°C
20	19.4
25	25.9
30	30.4
35	35.1
40	40.1
45	44.8

isn't as helpful as

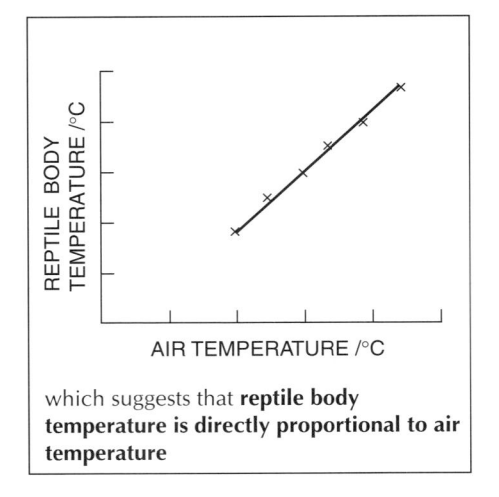

[Graph: REPTILE BODY TEMPERATURE /°C vs AIR TEMPERATURE /°C]

which suggests that **reptile body temperature is directly proportional to air temperature**

A graph may be produced from a table of data <u>following certain rules</u>

y axis

dependent (responding) variable

NB units separated from physical quantity by a solidus

e.g. **reptile temperature /°C**

or in words

e.g. **reptile temperature in °C**

There should be an **informative title:** simply **Effect of** (variable on *x* axis) **on** (variable on y axis).

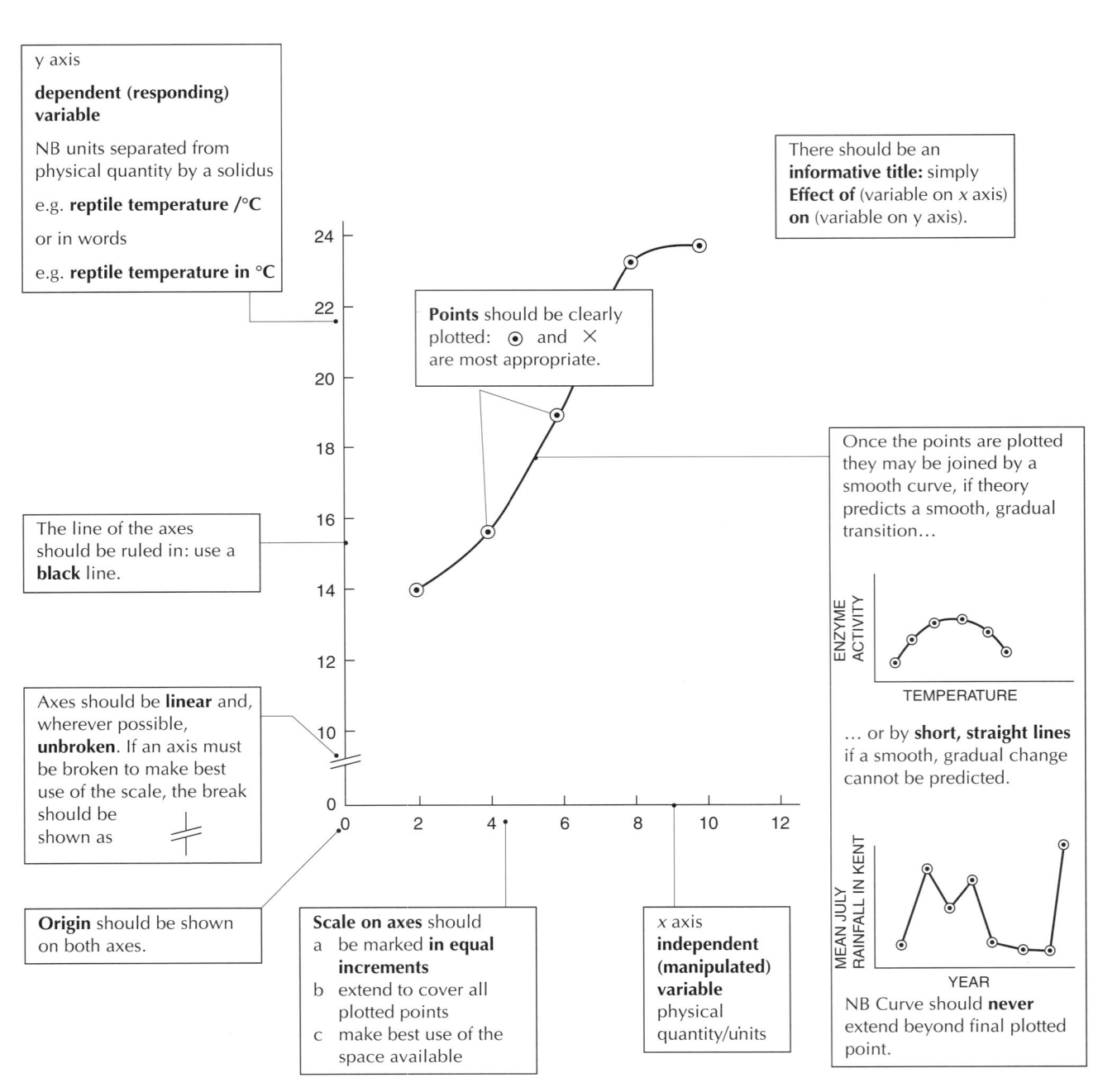

Points should be clearly plotted: ⊙ and × are most appropriate.

The line of the axes should be ruled in: use a **black** line.

Once the points are plotted they may be joined by a smooth curve, if theory predicts a smooth, gradual transition…

[Graph: ENZYME ACTIVITY vs TEMPERATURE]

… or by **short, straight lines** if a smooth, gradual change cannot be predicted.

Axes should be **linear** and, wherever possible, **unbroken**. If an axis must be broken to make best use of the scale, the break should be shown as ⟍⟋

Origin should be shown on both axes.

Scale on axes should
a be marked **in equal increments**
b extend to cover all plotted points
c make best use of the space available

x axis
independent (manipulated) variable
physical quantity/units

[Graph: MEAN JULY RAINFALL IN KENT vs YEAR]

NB Curve should **never** extend beyond final plotted point.

INDEX

absorption 13
accommodation 23
Acetobacter 53
acid rain 58
acne 31
active site 10
active transport 5, 9, 60, 64
adaptation 55
additives 17
adenine 43, 44
ADH (antidiuretic hormone) 22, 27
ADP (adenosine diphosphate) 5
adrenaline 22
Agrobacterium tumefasciens 49
AI (artificial insemination) 67
albinism 43
alcohol 20, 24, 51, 54
algae 40, 59, 61
allele 42, 44, 45
amino acid 11, 13, 26, 43, 64
ammonium ion 64
amnion 33
amniotic fluid 33
amylase 10, 13, 60
anaemia 11, 12
angina 14
Angiosperm 7, 40
Anopheles 16
anorexia 12
antagonistic pair 30
anther 38
antibiotic 16, 18, 20, 52, 65, 67
antibody 18, 19, 33
antibody, monoclonal 19
antigen 19
antiseptic 16, 60
aorta 14
aphid 35, 63, 66
artery, coronary 14
artery, hepatic 14
artery, pulmonary 14
artery, renal 14, 27
artery, umbilical 33
arthritis 30
arthropod 40
artificial selection 52
athlete's foot 18
atmosphere 58
ATP (adenosine triphosphate) 5, 30
atrium 14
autotroph 55
axes 70
axon 21

bacteriocide 18, 52
bacteriostatic 52
bacterium 18, 40, 48, 50, 51, 52, 53, 59, 60, 61, 64
baking 54
barbiturate 20
base pair 43, 44
bee 38
bicarbonate indicator 5

biceps 30
bile 13
biogas 51
biological control 63, 66
bioreactor 51, 53
bladder 27, 31
blind spot 23
blood cell, red 15, 30
blood cell, white 15, 30
blood 15
BOD (biological oxygen demand) 59, 61, 65
body temperature 5, 15, 29
brain 24, 32
bread 54
bronchitis 20
budding 54

caffeine 20
calcium 11, 15, 18
cambium 35
cancer, bladder 20
cancer, colon 12
cancer, lung 16
cancer, skin 16, 58
cap (diaphragm) 31
capillary 13, 14, 18
carbohydrate 11, 64
carbon cycle 60, 64
carbon dioxide 5, 15, 26, 30, 33, 34, 51, 54, 58, 64, 66
carnivore 55, 62
carrier proteins 9
carrier 45
carrying capacity 56, 57
cartilage 30
catalyst 10
cell division 5, 42, 43, 44
cell membrane 8, 9, 11, 13
cell, plant and animal 8
cellulose cell wall 8
cellulose 34
centromere 44
cerebellum 24
cerebral cortex 24
cervix 31, 33
CFC (chlorofluorocarbon) 58
cheese 53
chlorophyll 8, 34, 37
chloroplast 8
cholera 16, 17
cholesterol 11, 12
choroid 23
chromatid 42
chromosome 8, 42, 43, 44, 46, 48
ciliary muscle 23
cirrhosis 16, 20
clinistix 10, 25
clone 50
cobalt chloride paper 36
coil 31
colour blindness 46
combustion 64
community 55
condom 31
cone cell 23
constipation 12

consumer 62
contraception 31, 32
control 68
cornea 23
cotyledon 39
crossing over 47
cuticle 7, 36
cystic fibrosis 16, 45
cytoplasm 8
cytosine 43, 44

DDT 59
deamination 26
decomposition 62
denaturation 10, 29
denitrification 64
depression 16
diabetes 16, 25
dialysis 28
diaphragm 6
diarrhoea 17, 20
diet, adequate 11
diet, balanced 11
diffusion 9
digestion 13
diploid 42, 46
disease 16
dispersal - wind, water, animal 39
DNA replication 42, 43
DNA 43, 44, 48
dominant 45
double helix 44
Down's syndrome 44
drugs 20, 24
drugs, effect on synapse 21
dysentery 17

ecology 55
ecosystem 55, 65, 66, 67
egg 50
embryo 33, 39, 50
emphysema 20
endocrine gland 22
endonuclease 48
endosperm 7
endothermy 6, 13, 29
energy 5, 9, 11, 54, 62
environmental resistance 47, 56, 65, 66, 67
enzyme 5, 8, 10, 48, 53, 60, 64
enzymes - uses 10
epidermis 35
epididymis 31
eutrophication 59
excretion 26, 64
eye 23

factor-8 48, 49
fair testing 68
fermentation 54
fermenter 48, 51, 52, 53
fern 40
fertilisation 39, 42, 44, 46, 47, 50
fertiliser 59
fever 29

fibre 11
fibrin 18
fibrinogen 13, 15, 18
filament 38
fish farming 65
fish 40, 65
flaccidity 36
flower 7, 38
foetus 33
food chain 60, 62
food preservation 53
fossil fuel 58, 64
fruit 39
fungus 18, 40, 50, 60, 64, 66

gall bladder 13
gamete 31, 39, 42, 44, 46, 47
gasohol 51
gene probe 45
gene transfer 49
gene vector 48
gene 8, 42, 43, 44, 45, 48
genetic engineering 10, 25, 48, 66
genotype 44, 45
genus 41
geotropism 37
global warming 58
glucagon 13, 22, 25
glucose 5, 11, 13, 15, 25, 30, 34, 54
glycogen 8, 13, 25
gonorrhoea 18
Graafian follicle 32
graphs 70
greenhouse effect 58
grey and white matter 21
growth 5, 11, 33, 67
growth, plant 7
guanine 43, 44
gut 13

habitat 55
haemoglobin 15
haemophilia 46
haploid 42
hayfever 16
heart disease 12, 14
heart 14
heat 5, 30, 60
hepatitis 16, 19
herbicide 66
herbivore 55, 62, 63
heroin 20, 24
heterotroph 55
heterozygous 45, 46
homeostasis 6, 15, 25, 29
homologous pair 42, 44, 46, 47
homozygous 45, 46
hormone 8, 11, 15, 20, 22, 31, 32, 37
horticulture 66
human growth hormone 22, 48
humerus 30
humidity 36, 66
Huntington's chorea 45

71

hybridoma 19
hydrochloric acid 18
hyperglycaemia 25
hypoglycaemia 25
hypothalamus 24, 27
hypothermia 20, 29
hypothesis 68

immunity - active, passive, natural, artificial 19
immunity 16, 17
independent assortment 47
influenza 16, 18
insect 40
insecticide 16, 66
insulin 13, 15, 22, 25, 48
iris 23
iron 11
irradiation 17
IUD (intra-uterine device) 31

joint 30

karyotype 46
key 40
kidney 26, 27, 28
kingdom 41
kwashiorkor 11, 12

lacteal 13
lactic acid 30, 54
large intestine 13
lead 58, 59
leaf 7, 34
lens 23
ligament 30
ligase 48
light 34, 66, 67
limewater 5
limiting factor 34
Linnaeus 41
lipase 10, 13, 60
lipid 9, 11
liver 13, 16, 26
loop of Henle 27
louse 65
LSD 20, 24
lung 6, 14, 26
lymphocyte 15, 18, 19
lysozyme 18

magnesium 37
malnutrition 11, 12, 20
mammal 40
mammary gland 6
marasmus 11, 12
mean 68
medulla 24
meiosis 42, 44, 47
membrane, selectively-permeable 9
Mendel's law 45
meninges 24
meningitis 24
menstrual cycle 32
menstruation 32
meristem 50
mesophyll 34
metabolism 8, 13, 15
methane 51, 58, 60, 61
microorganism 51
micropyle 39
mineral ions 9, 11, 34, 37
mitochondria 8
mitosis 42
moss 40

mould 52
mouth 13
movement 5
muscle 30
mutagen 43
mutation 43, 44, 47
myelin sheath 21

natural selection 47
negative feedback 25, 27, 29, 32, 56
nephron (kidney tubule) 27
nervous system 6, 21
neurone 21
neurotransmitter 21
night blindness 12
nitrate 37, 59, 64
nitrification 64
nitrogen cycle 60, 64
nitrogen fixation 49, 64
nitrogen 37, 64
nitrogen-fixing bacteria 7
node of Ranvier 21
nucleotide 43, 44
nucleus 8, 43

obesity 12
oesophagus 13
oestrogen 22, 32
oil 59
optic nerve 23
osmosis 9
osteoporosis 12
ovary 7, 31, 38
oviduct 31
ovulation 32
ovule 39
oxygen debt 54
oxygen 15, 30, 33, 34, 54, 59, 60
ozone 58

painkiller 20
pancreas 13, 25
parasite 63
pasteurisation 17, 53
pathogen 15, 16, 18
penicillin 52
Penicillium 52
penis 31
pepsin 13
peristalsis 13
pest 63
pesticide 49, 59, 63
petal 38
pH 10, 15
phagocyte 8, 15, 18
phenotype 43, 44, 45
pheromone 63
phloem 34, 35
phosphate 59
photosynthesis 34, 62, 64
phototropism 37
pill 32
pituitary hormones 31, 32
placenta 33
plasma 15, 26
plasmid 48, 49
Plasmodium 16
platelets 15, 18
polio 19
pollen 39
pollination 38, 66
pollution 58, 59
population 55, 56, 57, 63
positive feedback 59

potometer 36
predator 63
prey 63
producer 55, 62
progesterone 22, 31, 32
prostate gland 31
protease 10, 60
protein 9, 11, 13, 18, 43, 44, 64, 65
protoctista 40
Pruteen 53
pulse 14
pupil 23
pyramid of biomass 62
pyramid of numbers 62
pyruvic acid 5, 54

radiation 43, 58
radiotherapy 16
radius 30
receptacle 38
recessive 45
rectum 13
reflex action 24
reflex arc 21
refrigeration 17
rennin 53
reptile 40
resistance 49, 52
respiration 13, 26, 62, 64
respiration, aerobic 5
respiration, anaerobic 30, 54
respirometer 5
restriction enzyme 10, 48
retina 23
Rhizobium 7, 49
ribosome 8, 43
rickets 11, 12
rod cell 23
root 7, 35
root nodule 7, 64
roughage 11, 12

Salmonella 16, 17
saprotroph 17, 60
schizophrenia 16
SCP (single cell protein) 53
scrotum 31
scurvy 11, 12, 16
seed 39
selective breeding 65, 66
sepal 38
sewage 59, 60, 61
sex linkage 46
sex 46
shivering 29
sickle cell anaemia 16, 43
skeleton 30
skin 6, 29
sludge 61
small intestine 13
smoking 14, 16
solvent abuse 20
species 41
sperm 31, 50
spermicide 31
sphincter 27
spider 40
spinal cord 21
sporulation 52
stamen 38
starch 8, 11, 13, 34
stem 35
sterilisation 17, 31, 50
stigma 38
stomach 13

stomata 7, 34, 36
style 38
sucrose 11, 34
surrogacy 50
sweating 29
synapse 21
synovial fluid 30

tables 69
tendon 30
testis 31
testosterone 22, 31
thymine 43, 44
tobacco 20
tooth decay 12
toxin 17
transcription 43
transgenic organism 49
translation 43
transpiration 36
transplant 28
triceps 30
turgidity 8
typhoid 17

ulna 30
umbilical cord 33
undernutrition 11
units 69
urea 15, 26, 27, 33, 64
ureter 27, 31
urethra 27
uterus 31, 32, 33

vaccine 19
vacuole 8
vacuole, contractile 8
vagina 31
valve 14
valve, cuspid 14
valve, semi-lunar 14
variables 68
variation 47
vas deferens 31
vascular bundle 35
vasectomy 31
vasoconstriction and vasodilation 29
vegetative propagation 50
vein, hepatic portal 14
  hepatic 14
  pulmonary 14
  renal 14, 27
  umbilical 33
vena cava 14
ventricle 14
vesicle 8
villus 13, 33
vinegar 51, 53
virus 18, 33
vitamins 11
vulva 31

water potential 9, 15
water 11, 15, 34, 36, 60, 64
weedkiller 37
wilting 36

X chromosome 46
xylem 7, 9, 34, 35, 36

Y chromosome 46
yeast 54
yellow spot (fovea) 23
yoghurt 53

zygote 33, 39, 42, 50